Subsurface Carbonate Depositional Models:
A Concise Review

Subsurface

Carbonate Depositional

Models:
A Concise Review

by *George B. Asquith*

The Petroleum Publishing Company
Tulsa, Oklahoma

Copyright © 1979
by The Petroleum Publishing Company
1421 South Sheridan Road, P. O. Box 1260
Tulsa, Oklahoma 74101

Library of Congress Catalog Card Number: 78-65859
International Standard Book Number: 0-87814-104-9
Printed in the United States of America

1   2   3   4   5   83   82   81   80   79

Contents

*A Mark of Schlumberger

Preface

Depositional models can assist the petroleum geologist in determining rock types, predicting reservoir extent, and identifying depositional environments. It is therefore surprising that subsurface carbonate models are not described more frequently in the literature. When the models are presented, illustrations of the log responses associated with the environments of deposition are usually not emphasized.

Modeling is important to the subsurface geologist because it provides a framework within which separate and seemingly unrelated data acquires significance. A model can be used as a standard of comparison and can help answer basic questions, such as reservoir geometry, trend, and thickness variability. A subsurface depositional model is constructed in three dimensions, using the elements of rock type, depositional sequence, geometry, trends, and all bounding lithologies. These elements are interrelated and have a general predictability in similar sequences.

A geologist, attempting to solve carbonate subsurface problems by the use of models, faces the onerous task of searching through the literature and piecing together information from various sources. Time and resource limitations often frustrate his efforts. This work was written in response to such frustrations, and it attempts to summarize the identifying criteria of six basic subsurface carbonate depositional models. The approach is to emphasize the general, rather than the specific, characteristics of each model. This type of presentation should serve as a useful review for geologists working with subsurface carbonates, as a reference point for specific problem solving, and as an overview for those who wish to learn about subsurface carbonate environments. The book is not intended as a definitive source on subsurface carbonate modeling but rather as a bridge to the available literature.

The book begins with an introduction to basic principles of model construction—the parameters of a model and how they are developed. Incorporated in the text is a review of carbonate rock classification, a necessary first step in modeling. Chapter 2 has been included because of the increasing emphasis on post-depositional change (diagenesis). It is written as a synopsis in order to promote understanding of an extremely complex subject. Chapters 3-6 develop six basic subsurface carbonate depositional models, using case histories as prototypes. The corresponding log responses are presented for each model. The book concludes with a discussion of the methods for determining carbonate rock types from various log responses and identifying carbonate depositional environments directly from logs. This is an area which standard text books often tend to ignore or treat superficially, but its value cannot be overemphasized, because more than 90 percent of the subsurface information used by the petroleum geologist is derived from logs.

Acknowledgments

The author would like to express his sincere appreciation to Alpar Resources of Perryton, Texas for providing financial aid and to Mr. R. L. Parker and Mr. C. R. Gibson of Alpar for their encouragement and critical review of the manuscript. The author is most grateful for the data provided by Mr. H. S. Robinson of Robinson Resources, Calgary Alberta, Canada and Mr. Frank Frazier of American Stratigraphic Company, Denver, Colorado. Mr. W. E. Barone of Oil Development Company of Texas, Amarillo, Texas was especially helpful with his suggestions on diagenesis and his help with photomicrographs.

The assistance given by Mr. T. D. Kirkpatrick of Anadarko Production Company, Denver, Colorado in reviewing the chapter on regressive shoreline grainstones is appreciated. Mr. J. M. Baker of Union Texas Petroleum, Midland, Texas thoughtfully gave his comments and suggestions. The author is grateful to Schlumberger Well Services of Houston, Texas for permission to use several of their logging charts.

The following companies provided logs and samples: Atlantic Richfield Company, Dallas, Texas, Amarillo Oil Company and Diamond Shamrock Company of Amarillo, Texas, J. M. Huber Company, Borger, Texas, and Shell Oil Company, Corpus Christi, Texas. Students in cartography classes at West Texas State University under the direction of Dr. Robert D. Sawvell spent many hours drafting figures. Dr. Sawvell's aid and suggestions on graphics is greatfully acknowledged.

The author is indebted to Dr. G. M. Friedman of Rensselaer Polytechnic Institute for offering some invaluable comments and suggestions for improving the manuscript, and to Ann L. Asquith, Dr. W. C. J. vanRensburg, and Dr. F. W. Daugherty for their help with preliminary editing.

1

An Introduction
to Carbonate Petrography
and Modeling

Introduction

Limestones and dolomites are an economically important group of sedimentary rocks and, along with sandstones, contain virtually all the worlds supply of petroleum (Ham and Pray, 1962, p. 2). In order to facilitate exploration for petroleum in carbonate reservoirs, subsurface geologists need to recognize carbonate rock types associated with different carbonate depositional models, and relate the carbonate rock types to log responses. The knowledge acquired from identifying and classifying rock types and subsequently constructing a carbonate depositional model will aid the subsurface geologist in both exploration and development drilling of carbonate reservoirs.

Carbonate rock classification, therefore, is an essential first step in model construction. It allows for the grouping of rocks with similar petrographic characteristics, and, once grouped, the rock types can be used to construct vertical sequences and lateral facies for a particular depositional model. The importance of classification to model construction dictates that we begin our discussion with some background on how carbonates are classified and the fundamental components of classification.

Carbonate Classification

GENERAL

A carbonate rock consists of 50 percent or more carbonate minerals, mainly calcite and dolomite. Most ancient limestones are composed of low-magnesium calcite, as opposed to recent carbonate muds and sands (Chave, 1954) which are composed of appreciable amounts of metastable calcium carbonate (i.e. aragonite and high-Mg calcite). If dolomite is the only carbonate mineral in the rock, it is called a dolomite or a dolostone. However, when dolomite content ranges between 0 and 50 percent, the rock is called a dolomitic limestone.

If dolomite content ranges from more than 50 percent but less than 100 percent, it is called a limey dolomite. The common non-carbonate minerals found in carbonate rocks include anhydrite, gypsum, glauconite, chert, quartz, and clay minerals. With increasing amounts of impurities, carbonates grade into terrigenous sands or muds or into chemical evaporites. In such cases, prefixes such as sandy, shaly, or anhydritic are added to the carbonate rock type.

Texture and composition are the two fundamental parameters used in all rock classifications. Texture means the size, shape, sorting, and arrangement of the rocks components. Unlike terrigenous sands and muds where composition is expressed mostly by mineralogy, carbonate classification stresses grain type (i.e. oolith versus brachiopod fragment) in making compositional distinctions (Horowitz and Potter, 1971, p. 4). Grain type is used because carbonates are commonly composed of only one or two minerals. Also, unlike terrigenous sands, the textural parameters of size and sorting in carbonates are not always related to the hydrodynamics of the environment.

Size is often controlled by the growth characteristics of a particular organism, and sorting can be controlled by the variety of organisms living in a particular environment and not by the hydrodynamics. In carbonates with diverse grain type, however, the degree of sorting and grain size can be used, as in terrigenous sands, to differentiate high and low energy environments.

Detrital mud, chemical precipitates (cement), framework components (detrital grains and in-situ organic structures), and pores are the fundamental building blocks of all sedimentary petrology (Horowitz and Potter, 1971). The goal of sedimentary petrography, then, is to give a unique geologic interpretation to these four rock components so that the depositional history of the rock can be interpreted. In order to more fully understand the four components, each must be carefully described.

Framework Components Discrete, detrital particles of sediment that form a framework are called allochems. They are mostly sand-size (0.0625 and 2.0 millimeters) but may also be silt-size. Table 1 summarizes the different types of grains (allochems) encountered in the analysis of carbonate rocks.

In-situ organic structures constitute another framework element

TABLE 1
VARIETIES OF CARBONATE ALLOCHEMS

INTRACLASTS—
fragments of penecontemporaneous, generally weakly consolidated carbonate sediment that have been eroded from the sea bottom and redeposited (Folk, 1962, p. 63).

DETRITIAL SKELETAL GRAINS—
crinoid, brachiopods, foraminifera, ostracads, etc. May be whole or fragmented. Commonly transported after death.

PELLETS (PELOIDS)—
a rounded or oval mass of structureless micrite (0.03 to 0.15 mm in diameter) that may be fecal in origin (Folk, 1962, p. 64).

OOLITES—
a spherical to ellipsoidal body 0.25 to 2.0 mm in diameter, which may or may not have a nucleus and has concentric or radial structure or both (Glossary of Geology and Related Sciences, 1962).

PISOLITE—
a spherical or subspherical, accretionary body over 2.0 mm in diameter (Glossary of Geology and Related Sciences, 1962).

and include corals, bryozoans, and algae. These form an important group of carbonates (boundstones) whose framework is held together by secreted calcium carbonate.

Lime Mud Folk (1959, p. 8) first described the microcrystalline calcite ooze that is now almost universally called micrite. Micrite is the carbonate analog to argillaceous mud sometimes present between grains in terrigenous sands.

Cement The pore system of a carbonate rock may be partially or completely filled with chemically precipitated cement. Calcite is the most common cement, but evaporite minerals (i.e. gypsum, anhydrite, halite) and dolomite are also found. In contrast to dark opaque micrite, carbonate cement is often clear and crystalline. Folk (1959, p. 8) evoked the term sparry calcite or sparite to describe carbonate cement. The different types of cement and their time and environment of precipitation will be discussed in the chapter "Introduction to Diagenesis."

Pores It is the pore system (porosity and permeability) that permits the movement of fluids (oil, gas, and water) through the rock. The formation and destruction of porosity, as well as its distribution, are of fundamental importance in the study of carbonate rocks. The different types of porosity and how they are formed and destroyed will be discussed later in the chapter on diagenesis.

DUNHAM CLASSIFICATION

The carbonate classification used throughout the text is called the "Classification of Carbonate Rocks According to Depositional Texture," and was proposed by Dunham of Shell Research in 1962. Table 2 illustrates the essential aspects of the Dunham classification. A major distinction in this classification is whether the rock is mud supported (grains not touching: mudstones and wackestones) or grain supported (grains in contact: packstones and grainstones). Grainstones lack the lime mud found in packstones, mudstones, and wackestones (Table 2). The presence or absence of lime mud (micrite) is believed by Dunham (1962) to be the most reliable indicator of wave energy.

Calm water allows mud to settle to the bottom and remain while agitated water does not. The degree of water agitation, therefore, can be determined in carbonate rocks by the ratio of micrite (lime mud) to sparry calcite cement. Friedman and Sanders (1978, p. 175), however, suggest caution should be used with this geologic interpretation because lime mud may filter into empty pores after deposition or a micritic matrix may result from the precipitation of crypto-crystalline calcite cement in a high energy grainstone. Also, micrite, if recrystallized (pseudospar), can resemble sparry calcite cement (see: chapter 2). If

TABLE 2
CLASSIFICATION OF CARBONATE ROCKS ACCORDING
TO DEPOSITIONAL TEXTURE
(MODIFIED AFTER DUNHAM, 1962, P. 108-121)

I. Mud-Supported Textures (allochems floating in mud matrix)
   A. LIME MUDSTONE—less than 10 percent allochems (example: ostracod mudstone).
   B. WACKESTONE—more than 10 percent allochems (example: crinoid-bryozoan wackestone).
II. Grain-Supported Textures (allochems form a self supporting framework)
   A. PACKSTONE—grain-supported with mud matrix between grains (example: peloid packstone).
   B. GRAINSTONE—lacks mud matrix and is grain-supported (example: oolite grainstone).
III. Original Components Bound Together During Deposition
   A. BOUNDSTONE—skeletal material intergrown or held together by encrusting organisms.
   (example: stromatoporoid boundstone).
IV. Depositional Texture Unrecognizable
   A. CRYSTALLINE CARBONATE—original texture destroyed by diagenesis
   (example: sucrosic dolomite).

these factors are not taken into account, errors in environmental interpretation will occur.

Dunham (1962) classifies carbonate rocks made up of in-situ organic structures as boundstones. If the texture of a carbonate rock is unrecognizable, the rock is then called a crystalline carbonate. When more detailed descriptions of carbonate rocks are needed, modifiers are added. For example, a rock can be classified as a brachiopod packstone or a shaly, pelecypod wackestone and so forth. A rock which contains more than one type of allochem is described according to the following rules: (1) the allochem in greatest abundance is mentioned last, and (2) the allochem types are separated with a hyphen (Friedman and Sanders, 1978, p. 172). It is understood, therefore, that a rock described as a crinoid-brachiopod wackestone has brachiopod fragments in excess of crinoid fragments.

The Dunham (1962) classification system does not include dolomites. Consequently, for the purposes of this discussion, they are described as follows: If the original limestone texture is preserved after dolomitization, the prefix "dolomitized" is used (e.g. dolomitized oolite grainstone). Detailed descriptions of dolomites include the crystal shape — euhedral, subhedral or anhedral. A fabric with predominantly euhedral crystals (i.e. well developed intercrystalline porosity) is called idiotopic or sucrosic. Continued growth of dolomite crystals causes the individual rhombs to grow together and fill pore space. The resulting fabric has subhedral crystals and is called hypidiotopic. The crystals may, with further growth, become anhedral and form a xenotopic fabric (Friedman and Sanders, 1978, p. 179).

When a carbonate has been classified, an environmental interpretation can be made and mapped. Other useful methods of carbonate classification are presented in the American Association of Petroleum Geologists Memoir 1, "Classification of Carbonate Rocks" edited by W. E. Ham (1962).

GENERAL CARBONATE MODELS AND CYCLES

Introduction to Modeling   A sedimentary model has been described by Potter and Pettijohn (1963, p. 226) as an intellectual construction of a depositional unit based on a prototype. A Bouma turbidite sequence, a point bar, a Cretaceous reef, or a carbonate basin can all be represented by models. A sedimentary model may be based on either an ancient or a modern example, or a combination of both. A model integrates the different components of a depositional unit such as rock type, geometry, and diagenesis, and helps the geologist identify interrelationships between these components. (Horowitz and Potter, 1971).

In carbonate models, unlike terrigenous clastic models, diagenesis is often as important to the distribution (geometry) of the porosity as the original facies. The major value of depositional models is that they provide a basis for prediction of lithologic distributions in three dimensions. The important parameters needed to establish a reliable carbonate depositional model are listed in Table 3.

Figure 1-1 shows two Cretaceous regressive carbonate models: a reef (Fig. 1-1A) and a tidal flat (Fig. 1-1B). Figure 1-1C illustrates the position of time lines, through sequences such as these, that are used to establish lateral carbonate facies. On the other hand, vertical lines drawn through these sequences (Fig. 1-1A and 1-1B), establish cycles (sequences) for the models at different positions within the basin. Once a geologist has determined the model type, then the depositional cycle encountered in a particular well establishes that well's position in the basin.

Figure 1-2 illustrates that the degree to which a geologist can predict a well's position within the basin from a depositional cycle

TABLE 3

DISTINGUISHING FEATURES OF CARBONATE MODELS

(MODIFIED AFTER: HOROWITZ AND POTTER, 1971, P. 12; PETTIJOHN, ET AL., 1973, P. 445)

GEOMETRY:
Elongate, tabular, pods, and mounds.

ORIENTATION:
Parallel or perpendicular to the paleoshoreline or depositional strike.

DIMENSIONS:
length, width, and thickness.

COMPOSITION:
Mineralogy, petrology, fossils, sedimentary structures, biogenetic structures, and pore system.

INTERNAL ORGANIZATION:
Distribution of carbonate rock types, sedimentary structures, biogenetic structures, texture, and pore system.

POSITION IN BASIN:
Lateral (margin versus center of basin) and relation to major paleophysiographic features such as shelf edge, paleoshoreline, etc.

BOUNDING LITHOLOGIES:
Lateral (seaward and landward equivalents), and vertical lithologies and contacts (gradational, sharp, or erosional).

DIAGENESIS:
The changes after deposition and their effect on the pore system.

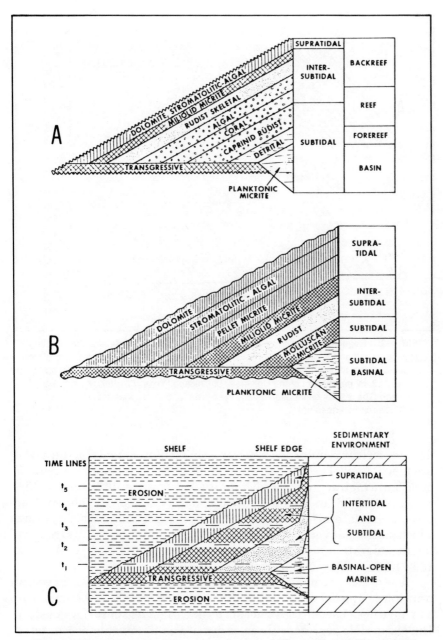

FIG. 1-1: Cretaceous reef (Fig. 1-1A) and non-reef (Fig. 1-1B) depositional cycles. Figure 1-1C illustrates the position of time lines through depositional cycles (Coogan, 1969, Figs. 7 and 11). From *Symposium on cyclic sedimentation in Permian Basin,* courtesy of the author and the West Texas Geological Society.

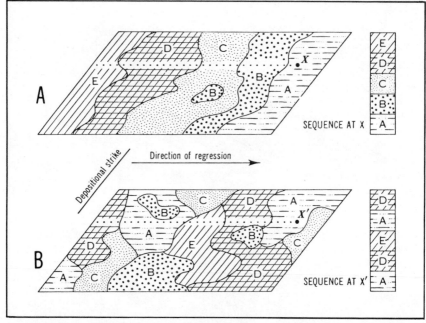

FIG. 1-2: Regular versus "crazy quilt" distribution of environments, and the resulting lithologies at X and X' produced by regression from left to right. Reprinted from Random processes and lithologic transitions by P. E. Potter and R. F. Blakely, Jour. of Geology, v. 76, p. 159, by permission of the University of Chicago Press. © 1968 by University of Chicago. The regular distribution results in an orderly vertical sequence and the "crazy quilt" distribution does not.

depends on how regular the different facies (environments) of the model are distributed. The regular example (Fig. 1-2A) results in a regular cycle and a high degree of predictability. The "crazy quilt" example (Fig. 1-2B) results in an irregular cycle and very low predictability. The prediction of lateral facies from vertical cycles is based on Walther's Law which states that vertical lithologic change in a section results from the lateral migration of different facies (environments) past it (Walther, 1892/1893, p. 62).

The discussion of models thus far has been concerned with general principles of model construction. These principles in subsequent chapters have been used to construct six basic carbonate subsurface models. The six models represent those carbonate environments encountered most frequently by the subsurface geologist. Both carbonate rock types and log responses of each model are presented.

REFERENCES

Chave, K. I., 1954, Aspects of the biogeochemistry of magnesium. 1. Calcareous marine organisms: *Jour. Geology,* v. 62, p. 266–283.

Coogan, A. H., 1969, Recent and ancient carbonate cyclic sequences *in* Elam, J. G., and Chubers, S., *eds., Symposium on cyclic sedimentation in Permian Basin:* Midland, Texas, West Texas Geological Soc., p. 5–16.

Dunham, R. J., 1962, Classification of carbonate rocks according to depositional texture, p. 108–121 *in* Ham, W.E., *ed., Classification of carbonate rocks, a symposium:* Am. Assoc. Petroleum Geologists, Mem. 1, 279 p.

Folk, R. L., 1959, Practical petrographic classification of limestones: *Am. Assoc. Petroleum Geologists, Bull.,* v. 43, p. 1–38.

———, 1962, Spectral subdivision of limestone types, p. 62–84 *in* Ham, W.E., *ed., Classification of carbonate rocks, a symposium:* Am. Assoc. Petroleum Geologists, Mem. 1, 279 p.

Friedman, G. M., and Sanders, J. E., 1978, *Principles of Sedimentology:* New York, John Wiley and Sons, 792 p.

Ham, W.E., and Pray, L. C., 1962, Modern concepts and classification of carbonate rocks, p. 2–19 *in* Ham, W.E., *ed., Classification of carbonate rocks, a symposium:* Am. Assoc. Petroleum Geologists, Mem. 1, 279 p.

Horowitz, A. S., and Potter, P. E., 1971, *Introductory Petrography of Fossils:* Berlin, Heidelberg, and New York, Springer-Verlag, 302 p.

Howell, J. V., *ed.,* 1962, *Glossary of Geology and Related Sciences:* 2nd ed., Washington, D.C., The Am. Geological Inst., 325 p.

Pettijohn, F. J., Potter, P. E., and Siever, R., 1973, *Sand and Sandstone:* New York, Heidelberg and Berlin, Spring-Verlag, 618 p.

Potter, P.E., and Pettijohn, F. J., 1963, *Paleocurrents and Basin Analysis:* Berlin, Heidelberg, and New York, Springer-Verlag, 296 p.

———, and Blakely, R. F., 1968, Random processes and lithologic transitions: *Jour. of Geology,* v. 76, p. 154–170.

Walther, Johnannes, 1892/1893, *Lithogenesis der Gegenwort:* Jena, Gustav Fischer, 1055 p.

SUGGESTED READING

Carbonate Classifiaction

Friedman, G. M., and Sanders, J. E., 1978, *Principles of Sedimentology:* New York, John Wiley and Sons, p. 167–194.

Ham, W. E., *ed.,* 1962, *Classification of carbonate rocks, a symposium:* Am. Assoc. Petroleum Geologists, Mem. 1, 279 p.

Scholle, P. A., 1978, *Carbonate rock constituents, textures, cements, and porosities:* Am. Assoc. Petroleum Geologists, Mem. 27, 241 p.

Carbonate Facies and Models

Friedman, G. M., *ed.,* 1969, *Depositional environments in carbonate rocks, a symposium:* Soc. Econ. Paleontologists and Mineralogists, Spec. Pub. No. 14, 209 p.

Purser, B. H., *ed.*, 1973, *The Persian Gulf, Holocene carbonate sedimentation and diagenesis in a shallow epicontinental sea:* Berlin, Heidelberg, and New York, Springer-Verlag, 471 p.

Wilson, J. L., 1975, *Carbonate Facies in Geologic History:* Berlin, Heidelberg, and New York, Springer-Verlag, 471 p.

2

Introduction
to Diagenesis

General

Carbonate reservoirs are characterized by extreme heterogeneity of porosity and permeability as a result of the complexity of original carbonate facies and later diagenetic influences (Jardine, et al., 1977, p. 873). These diagenetic influences either preserve and enhance original porosity or totally destroy it.

The importance of both the original carbonate facies and later diagenesis to the distribution of porosity can be illustrated by a study of a Lower Cretaceous (Edwards) patch reef in West Texas (Jacka and Brand, 1977). Jacka and Brand (1977, p. 379) concluded that virtually all primary porosity in the reef complex had been eliminated by lithification and cementation so that only secondary porosity remained. Also, in the same study, they noted that the development of leached (secondary) porosity was controlled by the distribution of aragonitic fossils (corals and caprinids) which were more abundant in the core, but less abundant in the forereef facies, causing a sharp decrease in porosity at the core-forereef boundary (Jacka and Brand, 1977, p. 379).

Jacka and Brand's study illustrates that even when virtually all porosity is secondary, the mapping of original carbonate facies can still be important because the original facies can control development of secondary porosity. In some cases, however, there are carbonate sequences, such as the Smackover Formation in Walker Creek Field in Lafayette and Columbia counties Arkansas, where porosity can be independent of the original depositional environment (Becher and Moore, 1976, p. 45).

Clarification of diagenetic terminology is necessary before undertaking a study of the subject:

Diagenesis—all processes that act on the rock after deposition but before the temperature and pressure become great enough to create metamorphic minerals.

Penecontemporaneous — changes that take place almost at the same time as deposition (i.e. changes before lithification).

Syngenetic—defined and used in the text as interchangeable with penecontemporaneous.

Neomorphism—all transformations within a mineral or a polymorph of that mineral, whether inversion or recrystallization, whether the new crystals are larger or smaller or simply differ in shape (i.e. textural change) from the previous ones (Folk, 1965, p. 21).

Pseudomorphic—neomorphism where there is an atom for atom replacement between minerals with similar atomic structures (example: calcite → Mg calcite). In this type of neomorphism, the original skeletal microstructure is preserved.

Xenomorphic—neomorphism where replacement is not on atomic scale and the minerals have different atomic structures (example: calcite → aragonite). In this type of replacement, the original skeletal microstructure is replaced by mosaic calcite and is destroyed (Friedman, 1975, p. 386).

Neomorphic—replacement or neomorphism (Folk, 1965, p. 21) whereby the new product has a different texture (Jacka, 1975, p. 38).

Paramorphic—replacement or neomorphism (Folk, 1965, p. 21) with no textural change (Jacka, 1975, p. 38).

Meniscus cement—fine (5-10μ), equant calcite cement concentrated at grain contacts and small pore spaces (Dunham, 1971).

Micrite rim cement—dense, dark, micrite coatings on allochem grains (Dunham, 1969).

Needle fiber cement—long, needle-like crystals of calcite cement. The needle fibers are 1-4μ wide and 125-200μ long (Ward, 1970, p. 121; Supko, 1971, p. 142).

Gravitational cement — cement that grows predominantly in a downward direction in response to gravity (Müller, 1971).

Eogenetic—the eogenetic stage of diagenesis applies to the time interval between final deposition and burial below the depth of significant influence by processes that either operate from the surface or depend for their effectiveness on proximity to the surface (Choquette and Pray, 1970).

Mesogenetic — the mesogenetic stage of diagenesis applies to the time interval during which rocks are buried below the major influence of the processes operating from or closely related to the surface (Choquette and Pray, 1970).

Telogenetic — the telogenetic stage of diagenesis applies to the time interval during which long-buried rocks are located near the surface again, as a result of crustal movement and erosion, and are significantly influenced by processes associated with the formation of an unconformity (Choquette and Pray, 1970).

A comprehensive explanation of carbonate textures and fabrics can be obtained in a review by Friedman (1965).

Understanding diagenetic processes is important in studying carbonate reservoirs because of their effect on both the formation and destruction of porosity (Murray, 1960, p. 59; Jardine et al., 1977). Table 1 is a listing of the different types of porosity classified by mode of origin (i.e. primary or secondary).

Figure 2-1 illustrates the different environments under which diagenesis takes place. Most carbonate diagenesis takes place during subaerial exposure because carbonate minerals in the subsea environment are in contact with pore water having the same composition as sea water (Blatt, et al., 1972, p. 469). Once pore water changes composition

TABLE 1
TYPES OF POROSITY

PRIMARY:
> **Intergranular** — pores between carbonate allochems in a non-cemented or partially cemented grainstone.
>
> **Organic**—pores developed within the skeleton of an organism (examples: the uncemented cells in a stromatoporoid) or holes between organisms.

SECONDARY:
> **Intercrystalline** — pores between carbonate crystals (examples: sucrosic dolomite and chalky limestone).
>
> **Vugular**—pores (vugs) created by leaching of the carbonate rock. If the vug is small, it is sometimes referred to as pin-point porosity.
>
> **Moldic porosity**—implies what was occupying the vug (i.e. fossilmoldic and oomoldic).
>
> **Fractures**—fissures or cracks in the carbonate rock. Fractures are favorable in that they increase both porosity and permeability, however they can be unfavorable in that they can cause channeling.

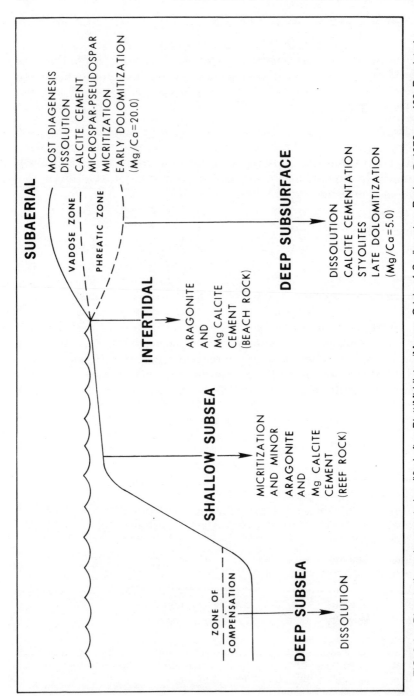

FIG. 2-1: Diagenetic environments (modified after: Blatt/Middleton/Murray, Origin of Sedimentary Rocks, © 1972, pg. 469. Reprinted by permission of Prentice-Hall, Inc., Englewood Cliffs, New Jersey.

(i.e. fresh H_2O; Fig. 1), and moves through the rock (vadose zone), dissolution (leaching) and cementation are possible. Beach rock (see: Fig. 2-1) results from cementation within the intertidal zone where the ocean spray penetrates porous beach grainstones and precipitates aragonite and Mg calcite by evaporation. Reef Rock (see: Fig. 2-1) forms by cementation in a shallow marine environment where aragonite and Mg calcite (Evamy, 1973, p. 330) are precipitated. The grain size of these submarine cements is microcrystalline with either a fibrous (aragonite) or a microsucrosic (Mg calcite) shape (MacIntyre, 1977, p. 503). MacIntyre (1977, p. 503) notes that submarine cementation is pronounced on the reef in areas of high agitation (reef crest), and is absent or poorly developed in deeper, forereef talus and backreef sediments.

Cementation

Figure 2-2 illustrates the different varieties of calcite cement (sparite). Mosaic cement (Fig. 2-2A) commonly occurs within drusy cement, and if it is found alone, it may represent a micrite matrix that has undergone neomorphism (pseudospar). The distinction between sparite and pseudospar is very important in petrographic analysis because both a wackestone or a packstone may be erroneously classified and mapped as a grainstone. Table 2 lists petrographic criteria used to differentiate sparite from pseudospar.

TABLE 2
CRITERIA FOR DIFFERENTIATING SPARITE FROM PSEUDOSPAR
(MODIFIED AFTER: FOLK, 1965, P. 43-45)

Sparite Cement	Pseudospar
1. Texture always grain supported, if grain supported it could be either sparite or pseudospar.	1. Texture may be either grain or mud-supported, if mud supported it is pseudospar.
2. Clear as glass, sparkles in hand samples.	2. Usually cloudy due to insoluble impurities.
3. Crystal size often varies over a small area.	3. Crystal size mainly uniform.
4. Crystals mainly stop sharply at grain boundaries with no penetration.	4. Crystals often penetrate grains and contacts are gradational.
5. Crystal boundaries often straight.	5. Crystal boundaries often curved.
6. Crystal next to grain often perpendicular.	6. Crystals next to grain often random.

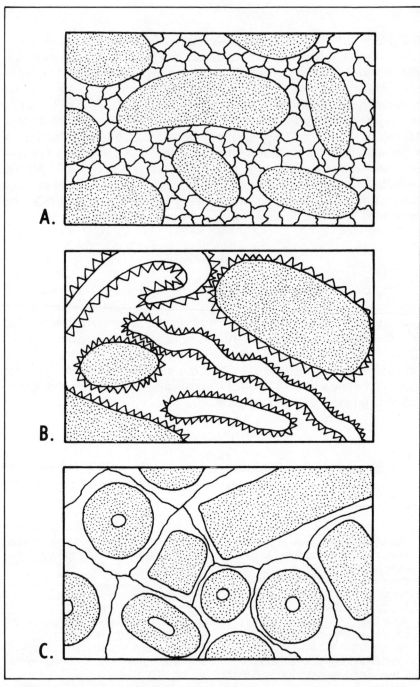

FIG. 2-2: Types of carbonate cement
A. Mosaic cement   B. Drusy cement   C. Epitaxial cement

Drusy cement (Fig. 2-2B) is, normally, the first layer of cement on all allochems except the echinoderms. It consists of fibrous or bladed crystals of calcite formed during subaerial exposure (Fig. 2-1).

Rim cement or epitaxial overgrowth (Fig. 2-2C) is produced when single crystals of calcite grow in optical continuity from the surface of echinoderm skeletal fragments. This type of cement often rapidly fills the pores (Jacka, 1974) of originally porous echinoderm grainstones. The cementation causes originally porous grainstones to become non-porous with the only porosity developed in associated packstones and wackestones either by leaching or dolomitization.

The Andrews South Devonian Field, Andrews County, Texas is an example of this type of diagenetic history (Lucia, 1962). Because of the problems caused by cementation, effort needs to be directed toward looking for lower energy micritic carbonates (packstones and wacke-stones) instead of grainstones. Such an approach is opposite the one used by geologists exploring for terrigeneous clastic reservoirs where reservoirs are predominantly developed in sandstones.

Neomorphism

Most skeletal material is made up of metastable forms of calcium carbonate (Mg calcite and aragonite). These change with time to stable, low Mg calcite (referred to simply as calcite) either with or without a change in pore water. Figure 2-3 illustrates: (1) the different types of neomorphism (taking into account original skeletal mineralogy) that skeletal debris go through under conditions of fresh water diagenesis, and (2) the different characteristics of the associated vadose and phreatic cements. Figure 2-4 illustrates the distribution of skeletal mineralogy both by geologic age and type of organism.

By combining Figures 2-3 and 2-4, it is possible to predict how the skeletal material of a particular organism will react to diagenesis, and, therefore, how it may appear in thin section. In addition to neomorph-ism, skeletal debris can undergo solution (leaching) and cementation (Fig. 2-3). The cement may be calcite, but dolomite and anhydrite cements are common. Table 3 lists the relative solubilities of carbonate minerals (arranged by decreasing solubility) as a function of: (1) mineralogy, (2) grain size, and (3) mineral shape. This table can also be used to predict the susceptibility of a carbonate mineral to replacement (i.e. dolomitization, silicification, etc.).

Lime mud, like the majority of skeletal debris, is composed mainly of metastable forms of calcium carbonate. Folk (1965) reported that lime mud, at the time of deposition, is in the form of small ($0.3\mu \times 0.3\mu \times$

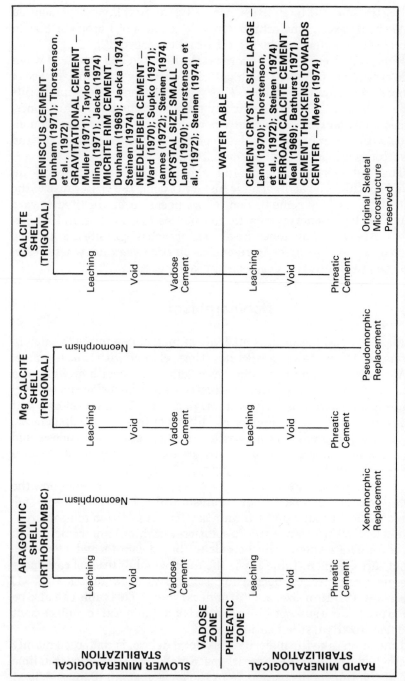

FIG. 2-3: Different types of neomorphism of skeletal debris under conditions of fresh water diagenesis, and varieties of vadose and phreatic cement. The susceptibilities of the different forms of calcium carbonate to leaching are listed in Table 3.

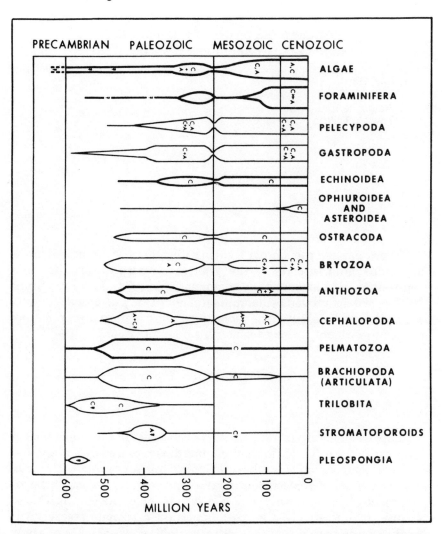

PRECAMBRIAN   PALEOZOIC   MESOZOIC   CENOZOIC

ALGAE

FORAMINIFERA

PELECYPODA

GASTROPODA

ECHINOIDEA

OPHIUROIDEA
AND
ASTEROIDEA

OSTRACODA

BRYOZOA

ANTHOZOA

CEPHALOPODA

PELMATOZOA

BRACHIOPODA
(ARTICULATA)

TRILOBITA

STROMATOPOROIDS

PLEOSPONGIA

600   500   400   300   200   100   0

MILLION YEARS

FIG 2-4: Distribution of carbonate secreting organisms through time. Reprinted from Biologic problems relating to the composition and diagenesis of sediments by H. A. Lowenstam *in* Earth Sciences: problems and progress in current research, *ed.* T. W. Donnelly, p. 174, by permission of the University of Chicago Press. © 1963 by William Marsh Rice University. C = calcite   A = Aragonite; Heavy lines = Mg Calcite.

3.0μ), elongate crystals; during neomorphism of lime mud to micrite, these crystals become small (1-4μ), blocky crystals. Micrite can undergo further neomorphism to form microspar (5-30μ, blocky crystals) or pseudospar ($>30\mu$, blocky crystals).

TABLE 3
SOLUBILITES OF CALCIUM CARBONATE
(ARRANGED BY DECREASING SOLUBILITES)

MINERALOGY
1. High Mg Calcite (>4% $MgCO_3$)
2. Aragonite
3. Low Mg Calcite (<4% $MgCO_3$)
SIZE
1. Fine
2. Coarse
SHAPE
1. Elongate
2. Blocky

Microspar and pseudospar form in a diagenetic environment low in Mg^{2+} ions, and micrite is the stable neomorphic product in a magnesium-rich diagenetic environment (Folk, 1974). Jacka and Brand (1977, p. 380) found that neomorphism of micrite to pseudospar occurred in the phreatic zone, but in the vadose zone, most of the matrix remained as micrite with only minor neomorphism to microspar.

Replacement

GENERAL

Replacement differs from neomorphism as defined by Folk (1965, p. 21) because the original carbonate mineral is replaced by a mineral of different composition (i.e. silica, dolomite, anhydrite, etc.). The most important type of replacement to the petroleum geologist is the replacement of calcium carbonate ($CaCO_3$) by dolomite ($CaMg(CO_3)_2$), because, quantitatively, the most important carbonate reservoirs in North America are dolomites (Murray, 1960, p. 66–67). There are at least two types of dolomite; their petrographic characteristics, associated rock types, and diagenetic environments will be discussed separately.

SYNGENETIC DOLOMITE

In restricted environments (i.e. lagoons, salt lakes, and sabkhas), evaporation causes the precipitation of gypsum ($CaSO_4 \cdot 2H_2O$) or anhydrite ($CaSO_4$) which increases the Mg/Ca ratio above 5.0 (normal sea water). When this ratio is increased above 10.0, the following reaction takes place $(2CaCO_3 + Mg^{2+} \rightarrow CaMg(CO_3)_2 + 2Ca^{2+})$ as the water comes in contact with calcium carbonate sediment (Illing, et al., 1965, p. 101; Butler, 1969). The dolomite formed by this process is commonly referred to as syngenetic, penecontemporaneous, evaporative, or supratidal (sabkha) dolomite.

There are four dolomitization models for the movement of the Mg^{2+} rich brine that causes this type of dolomitization:

1. The seepage reflux model (Deffeyes, et al., 1965) postulates downward descending supersaline brines.

2. The capillary fringe model (Hsü and Schneider, 1973, p. 416) postulates ascending vadose water by capillary action during evaporation.

3. The evaporative pumping of sea water model (Hsü and Schneider, 1973, p. 317) postulates ascending Mg^{2+} concentrated sea water.

4. The evaporative pumping of ground water model (Hsü and Schneider, 1973, p. 317) postulates ascending Mg^{2+} concentrated ground water.

Hsü and Schneider (1973, p. 421), as a result of their work in the Abu Dhabi sabkhas in the Persian Gulf, suggest that evaporative pumping from a ground water source is the most satisfactory of the four models for explaining the movement of Mg^{2+} rich brines. Syngenetic dolomites have also been found in the Bahamas (Shinn, et al., 1965) and in the Netherland Antilles (Deffeyes, et al., 1965). Table 4 lists the different criteria used to distinguish syngenetic dolomites from diagenetic dolomites.

DIAGENETIC DOLOMITES

The criteria for recognizing diagenetic dolomites (Table 4) are better documented than is the origin of diagenetic dolomite. This is unfortunate since these later diagenetic dolomites normally develop better porosity (see: Table 4). There are three hypotheses to explain the origin of diagenetic dolomites:

1. Reaction between a fresh water lens and sea water "Dorag Model" (Badiozamani, 1973).

2. Reaction between interstitual water and limestone at elevated temperature and pressure (Blatt, et al., 1972, p. 483–484).

3. Influx of Mg^{2+} rich meteoric waters (hydrodynamic dolomitization; Jacka and Franco, 1975; Barone, 1976).

In the Dorag Model (Badiozamani, 1973), the carbonate sediment

TABLE 4
CRITERIA FOR DISTINGUISHING SYNGENETIC
AND DIAGENETIC DOLOMITES
(IN PART AFTER: FISHER AND RODDA, 1969)

	Syngenetic	Diagenetic
Grain size	very fine <10μ	fine to coarse 20-100μ
Fabric	xenotopic to idiotopic (sucrosic)	xenotopic to idiotopic (sucrosic)
Associated rocks	evaporites	marine carbonates
Fossils	restricted fauna	open marine fauna
Sedimentary structures	mud cracks stromatolites	bioturbation, cross-stratification
Porosity	low	high
Type of replacement	very selective; matrix only	not as selective; matrix and fossils

and the interstitual connate water must come in contact with fresh water for the reaction to take place between Mg^{2+} and the calcium carbonate. The model is satisfactory for explaining why structural highs are often selectively dolomitized (i.e. due to their higher topography, they more readily come in contact with fresh water). This mechanism appeared plausible until Steinen's work (1974) in the Barbados. In a series of cores taken through a fresh water lens, he found stabilized calcite above and aragonitic carbonates below the lens, but no dolomite.

Blatt, et al. (1972, p. 483–484) note that the Mg/Ca ratio of sea water (5.2) is chemically adequate to dolomitize calcium carbonate at surface temperature and pressure, yet calcite is stable under these conditions. They suggest that with burial, elevated temperatures and pressures are sufficient to drive the reaction ($2CaCO_3 + Mg^{2+} \rightarrow CaMg(CO_3)_2 + 2Ca^{2+}$) to the right.

In the hydrodynamic dolomitization model (Jacka and Franco, 1975; Barone, 1976), the diagenetic dolomite results from influx of magnesium-rich meteoric ground water at a shallow depth soon after burial. The meteoric ground water becomes enriched in magnesium as it moves through the magnesium-rich sabkha (Jacka and Franco, 1975; Barone, 1976). The intertidal and, sometimes, the upper subtidal facies are dolomitized to nonporous microcrystalline dolomites due to multiple nucleation throughout the rock from the high concentration of magnesium. As the water moves further into the subtidal facies, nucleation is more isolated because of the lower concentration of magnesium; therefore, porous sucrosic dolomite is formed. The author believes that the hydrodynamic dolomitization model as proposed by Jacka and Franco (1975) is the best model presently available to explain the diagenetic dolomite of sabkhas and their associated facies. Examples of this are coastal tidal flats (Chapter 4) and reefs (Chapter 6) which have sabkhas present in the backreef facies. There are, however, dolomites that appear to be unassociated with sabkhas, and their origin may be related to a reaction at higher temperature and pressure (Blatt, et al., 1972, p. 483) or perhaps to the Dorag model (Badiozamani, 1973).

DOLOMITE POROSITY

Dolomite is slightly denser (more compact crystal lattice) than calcite so that the reaction of calcite to dolomite yields some 12-13 percent intercrystalline (sucrosic) porosity (Murray, 1960, p. 73). The resulting dolomite should have a porosity, assuming no compaction during or after dolomitization, which is equal to the sum of the original porosity plus an increase of 12 to 13 percent of the original calcite present (Murray, 1960, p. 73). In order for the resulting dolomite to have porosity, the excess calcium ($2Ca^{2+}$) created by the reaction $Mg^{2+} + 2CaCO_3 \rightarrow CaMg(CO_3)_2 + 2Ca^{2+}$, must be removed.

Murray (1960, p. 73) has outlined two models for removal of the excess calcium: (1) local source dolomitization and (2) distant source dolomitization. In local source dolomitization, there is insufficient carbonate (CO_2, HCO_3^-, CO_3^{2-}) available, and so calcite dissolution accompanies dolomitization, thus forming and redistributing porosity. In distant source dolomitization, sufficient carbonate is brought to the site of dolomitization so that porosity destruction occurs. Only later calcite dissolution (subaerial exposure and leaching) can produce dolomite porosity.

A different theory for the origin of porosity in dolomite has been suggested by Friedman (1965). Friedman believes that idiotopic or sucrosic fabric is formed by the growth of randomly oriented, uniformly-sized dolomite rhombs; porosity develops with the removal of unreplaced calcite by dissolution.

REFERENCES

Badiozamani, Khorsrow, 1973, The Dorag dolomitization model — application to the middle Ordovician of Wisconsin: *Jour. Sedimentary Petrology,* v. 43, no. 4, p. 965–984.

Barone, W. E., 1976, Depositional Environment and Diagenesis of the Lower San Andres Formation: M.S. thesis, Texas Tech Univ., 93 p.

Bathurst, R.G.C., 1971, *Carbonate sediments and their diagenesis, Developments in Sedimentology 12:* Amsterdam, London, and New York, Elsevier Pub. Co., 620 p.

Becher, J. W., and Moore, C. H., 1976, The Walker Creek Field, a Smackover diagenetic trap: *Transactions Gulf Coast Assoc. Geol. Soc.,* v. 26, p. 34–56.

Blatt, H., Middleton, G. V., and Murray, R. C., 1972, *Origin of Sedimentary Rocks:* Englewood Cliffs, New Jersey, Prentice-Hall, 634 p.

Butler, G. P., 1969, Modern evaporite deposition and geochemistry of coexisting brines, the sabkha, Trucial Coast, Arabian Gulf: *Jour. Sedimentary Petrology,* v. 39, no. 1, p. 70–89.

Choquette, P. W., and Pray, L. C., 1970, Geological nomenclature and classification of porosity in sedimentary carbonates: *Am. Assoc. Petroleum Geologists, Bull.,* v. 54, p. 207–250.

Deffeyes, K. S., Lucia, F. J., and Weyl, P. K., 1965, Dolomitization of Recent and Plio-Pleistocene sediments by marine evaporite waters on Bonaire, Netherlands, Antilles, p. 71–88 *in* Pray, L. C., and Murray, R. C., *eds., Dolomitization and limestone diagenesis, a symposium:* Soc. of Econ. Paleontologists and Mineralogists, Spec. Pub. No. 13, 180 p.

Dunham, R. J., 1969, Vadose pisolite in the Capitan reef (Permian), New Mexico and Texas, p. 182–191 *in* Friedman, G. M., *ed., Depositional environments in carbonate rocks, a symposium:* Soc. of Econ. Paleontologists and Mineralogists, Spec. Pub. No. 14, 209 p.

————————1971, Meniscus cement, p. 297–300 *in* Bricker, O. P., *ed., Carbonate cements:* Baltimore and London, The John Hopkins Univ. Press, 376 p.

Evamy, B. D., 1973, The precipitation of aragonite and its alteration to calcite in the Trucial coast of the Persian Gulf, p. 329–341 *in* Purser, B. H., *ed., The Persian Gulf, Holocene carbonate sedimentation and diagenesis in a shallow epicontinental sea:* Berlin, Heidelberg, and New York, Springer-Verlag, 471 p.

Fisher, W. L., and Rodda, P. U., 1969, Edwards Formation (Lower Cretaceous), Texas, Dolomitization in a carbonate platform system: *Am. Assoc. Petroleum Geologists, Bull.,* v. 53, p. 55–72.

Folk, R. L., 1965, Some aspects of recrystallization in ancient limestones, p. 14–48 *in* Pray, L. C., and Murray, R. C., *eds., Dolomitization and limestone diagenesis, a symposium:* Soc. of Econ. Paleontologists and Mineralogists, Spec. Pub. No. 13, 180 p.

————————, 1974, The natural history of crystalline calcium carbonate, effect of magnesium content and salinity: *Jour. Sedimentary Petrology,* v. 44, p. 40–53.

Friedman, G. M., 1965, Terminology of crystallization textures and fabrics in sedimentary rocks: *Jour. Sedimentary Petrology,* v. 35, no. 3, p. 643–655.

——————————, 1975, The making and unmaking of limestones or the downs and ups of porosity: *Jour. Sedimentary Petrology,* v. 45, no. 2, p. 379–398.

Hsü, K. J., and Schneider, J., 1973, Progress report on dolomitization-hydrology of Abu Dhabi Sabkhas, Arabian Gulf, p. 409–422 *in* Purser, B. H., *ed., The Persian Gulf, Holocene carbonate sedimentation and diagenesis in a shallow epicontinental sea:* Berlin, Heidelberg, and New York, Springer-Verlag, 471 p.

Illing, L. V., Wells, A. J., and Taylor, J. C. M., 1965, Penecontemporary dolomite in the Persian Gulf, p. 89–111 *in* Pray, L. C., and Murray, R. C., *eds., Dolomitization and limestone diagenesis, a symposium:* Soc. of Econ. Paleontologists and Mineralogists, Spec. Pub. No. 13, 180 p.

Jacka, A. D., 1974, Differential cementation of a Pleistocene carbonate fanglomerate, Guadalupe Mountains: *Jour. Sedimentary Petrology,* v. 44, no. 1, p. 85–92.

——————————, 1975, Observations on paramorphic and neomorphic dolostones and associated porosity relations: (abstract), *Program Bull.* ann. meeting AAPG-SEPM, p. 38–39.

——————————, and Franco, L. A., 1975, Deposition and diagenesis of Permian evaporites, and associated carbonates and clastics on shelf areas of the Permian Basin *in 4th Symposium on Salt:* Cleveland, Ohio, Northern Ohio Geological Society, p. 67–89.

——————————, and Brand, J. P., 1977, Biofacies and development and differential occlusion of porosity in a Lower Cretaceous (Edwards) reef: *Jour. Sedimentary Petrology,* v. 47, no. 1, p. 366–381.

James, N. P., 1972, Holocene and Pleistocene and calcareous crust (caliche) profiles, criteria for subaerial exposure: *Jour. Sedimentary Petrology,* v. 42, no. 4, p. 817–836.

Jardine, D. E., Andrews, D. P., Wishart, J. W., and Young, J. W., 1977, Distribution and continuity of carbonate reservoirs, *Jour. Petroleum Technology,* no. 10, p. 873–885.

Land, L. S., 1970, Phreatic versus vadose meteoric diagenesis of limestones, evidence from a fossil water table: *Sedimentology,* v. 14, p. 175–185.

Lowenstam, H. A., 1963, Biologic problems relating to the composition and diagenesis of sediments, p. 137–195 *in* Donnely, T. W., *ed., The Earth Sciences, problems and process in current research:* Chicago, Ill., Univ. of Chicago Press, 195 p.

Lucia, F. J., 1962, Diagenesis of a crinoidal sediment: *Jour. Sedimentary Petrology,* v. 32, p. 848–865.

MacIntyre, I. G., 1977, Distribution of submarine cements in a modern Caribbean fringing reef, Galeta Point, Panama: *Jour. Sedimentary Petrology,* v. 47, no. 3, p. 503–516.

Meyer, W. J., 1974, Carbonate cement stratigraphy of the Lake Valley Formation (Mississippian) Sacramento Mountains, New Mexico: *Jour. Sedimentary Petrology,* v. 44, no. 3, p. 837–861.

Müller, G., 1971, Gravitational cement, an indicator of the subaerial diagenetic environment, p. 32–35 *in* Bricker, O. P., *ed., Carbonate cements:* Baltimore and London, The John Hopkins Univ. Press, 376 p.

Murray, R. C., 1960, Origin of porosity in carbonate rocks: *Jour. Sedimentary Petrology,* v. 30, p. 59–84.

Neal, W. J., 1969, Diagenesis and dolomitization of a limestone (Pennsylvanian of Missouri) as revealed by staining: *Jour. Sedimentary Petrology*, v. 38, p. 1040–1045.
Shinn, E. A., Ginsburg, R. N., and Lloyd, R. M., 1965, Recent supratidal dolomite from Andros Island, Bahamas, p. 112–123 *in* Pray, L. C., and Murray, R. C., eds., *Dolomitization and limestone diagenesis, a symposium:* Soc. of Econ. Paleontologists and Mineralogists, Spec. Pub. No. 13, 180 p.
Steinen, R. P., 1974, Phreatic and vadose diagenetic modification of Pleistocene limestone, petrographic observations from subsurface of Barbados, West Indies: *Am. Assoc. Petroleum Geologists*, v. 58, no. 6, p. 1008–1025.
Supko, P. R., 1971, "Whisker" crystal cement in Bahamian rock, p. 143–146 *in* Bricker, O. P., ed., *Carbonate Cements:* Baltimore and London, The John Hopkins Univ. Press, 376 p.
Taylor, J. C. M., and Illing, L. V., 1971, Variation in Recent beachrock cements, Qatar Peninsula, Persian Gulf, p. 32–35 *in* Bricker, O. P., ed., *Carbonate cements:* Baltimore and London, The John Hopkins Univ. Press, 376 p.
Thorstenson, D. C., Mackenzie, F. T., and Ristvet, B. L., 1972, Experimental vadose and phreatic cementation of skeletal carbonate sand: *Jour. Sedimentary Petrology*, v. 42, no. 1, p. 162–167.
Ward, W. C., 1970, Diagenesis of Quaternary eolianites of N.E. Quintana Roo, Mexico, *Ph.D. diss.*, Rice Univ., 121 p.

SUGGESTED READING

Diagenesis
Braunstein, Jules, 1972, Carbonate Rocks II: Porosity and Classification of Reservoir Rocks, Selected papers reprinted from *Am. Assoc. Petroleum Geologists Bull.*, Reprint Series No. 5, 197 p.
Choquette, P. W., and Pray, L.C., 1970, Geologic nomenclature and classification of porosity in sedimentary carbonates: *Am. Assoc. Petroleum Geologists Bull.*, v. 54, no. 2, p. 207–250.
Dott, R. H. Sr., 1972, Carbonate Rocks I: Classification-Dolomite-Dolomitization, selected papers reprinted from *Am. Assoc. Petroleum Geologists Bull.*, Reprint Series No. 4, 237 p.
Folk, R. L., and Land, L. S., 1974, Mg/Ca ratio and salinity; two controls over crystallization of dolomite: *Am. Assoc. of Petroleum Geologists Bull.*, v. 59, no. 1, p. 60–68.
Friedman, G. M., and Sanders, J. E., 1967, Origin and Occurrence of dolostones, p. 267–348 *in* Chilinger, G. V., Bissell, H. J., and Fairbridge, R. W., eds. *Carbonate Rocks, origin, occurrence, and classification:* Amsterdam, Elsevier Pub. Co., 471 p.
——————————, 1975, The making and unmaking of limestones or the downs and ups of porosity: *Jour. Sedimentary Petrology*, v. 45, no. 2, p. 379–398.
Murray, R. C., 1960, Origin of porosity in carbonate rocks: *Jour. Sedimentary Petrology*, v. 30, no. 1, p. 59–84.
Pray, L. C., and Murray, R. C. eds., 1965, *Dolomitization and Limestone Diagenesis, A symposium:* Soc. of Econ. Paleontologists and Mineralogists Spec. Pub. No. 13, 180 p.
Scholle, P. A., 1978, *Carbonate Rock Constituents, Textures, Cements, and Porosities:* Am. Assoc. Petroleum Geologists, Mem. 27, 241 p.

3

Regressive Carbonate
Shoreline Model

General

Chapter 3 and Chapter 4 are concerned with shoreline deposits where the quantity of terrigenous debris is so small that carbonate shoreline facies develop. Like their terrigenous clastic counterparts (Curray, 1969), these carbonate shoreline deposits prograde seaward (Fig. 3-1) during times of either falling or stable sea level (Evans, et al., 1973). Progradation occurs when sediment accumulation exceeds redistribution by waves and tides. When regression occurs as a continuous long term event, the barrier island (shoreline) and offshore facies are overlain by lagoonal and tidal flat deposits (Fig. 3-1).

This is important to stratigraphic entrapment of hydrocarbons, because the nonporous, evaporitic, tidal flat sediments are placed lateral to and overlying the shorezone grainstones and lagoonal wackestones (Fig. 3-1). Also, as will be discussed later in Chapter 4, tidal flat facies can be a source of magnesium-rich waters which dolomitize the underlying marine sediments to create secondary sucrosic porosity. If the regressive cycle is interrupted by a transgression, the shorezone grainstones are overlain by offshore marine wackestones.

The barrier island grainstone facies (Fig. 3-1) are predominantly composed of well-sorted skeletal debris and peloids, but in some barrier island systems, oolites are also present (Loreau and Purser, 1973). Loreau and Purser (1973, p. 283) reported that the oolites in these barrier island facies do not form in the beach foreshore environment, but are dispersed from the adjacent tidal channel facies (Fig. 3-1) by longshore currents. Evans, et al. (1973, p. 233–277) and Reineck and

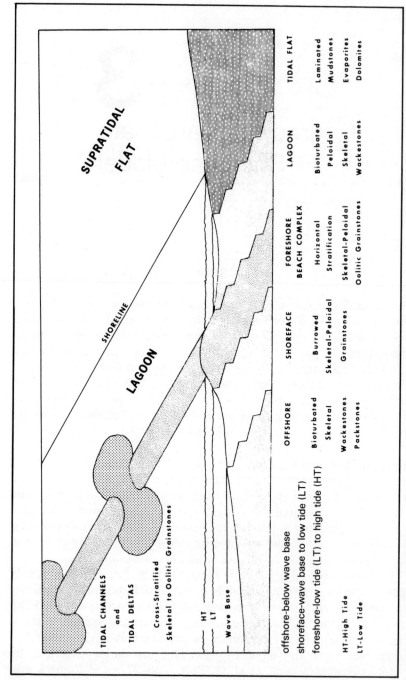

FIG. 3-1: Map and cross-sectional view illustrating the depositional environments and carbonate rock types associated with a regressive carbonate barrier island.

Singh (1975 p. 280–305) present thorough reviews of the sedimentation of modern carbonate and terrigenous clastic shoreline deposits.

Cretaceous Regressive Shoreline Grainstones, Nolan County, Texas

INTRODUCTION

On highway 70, approximately seven miles south of Sweetwater, Texas, there is an excellent exposure of the Lower Cretaceous Walnut-Comanche Peak Formation. The carbonates present at this outcrop represent shoreline deposits that developed adjacent to the Concho Arch (Moore, 1969). Figure 3-2 is a partial section illustrating the depositional sequence of this regressive carbonate shoreline. The cross-stratified, tidal channel grainstones near the top of the section fill scours cut into the beach and foreshore grainstones. The bioturbated, argillaceous, algal-mollusk wackestones overlying the sequence were deposited during a transgressive cycle which interrupted the shoreline regression (Fig. 3-2).

PETROGRAPHY

General The carbonates in this section (Fig. 3-2) can be subdivided into five rock types: (1) bioturbated, argillaceous, algal-mollusk wakestones; (2) horizontally stratified, burrowed, algal-echinoid grainstones; (3) low angle cross-stratified, algal-echinoid grainstones; (4) high angle cross-stratified, mollusk grainstones; and (5) bioturbated, algal-echinoid wackestones. The distribution of these different carbonate rock types is illustrated in Figure 3-2.

Detailed Descriptions *Bioturbated, argillaceous, algal-mollusk wackestone.*—codiacean green algae and pelecypod fragments in a medium-gray, argillaceous, micritic matrix (Figs. 3-3 and 3-4). Additional skeletal debris includes minor gastropods, echinoid spines, and miliolids.

Horizontally stratified, algal-echinoid grainstone. — light-gray, well-sorted, medium-grained (0.25 to 0.5 mm) fragments of green algal plates and echinoid fragments with epitaxial cement (Fig. 3-5). Additional allochems are peloids and pelecypod fragments. Vertical burrows of the sub-shoreface, shoreface filter feeder "ophimorpha" commonly occur.

Low angle cross-stratified, algal-echinoid grainstone.—tan, well-sorted, coarse-grained (0.5 to 1.0 mm) fragments of codiacean green algae and

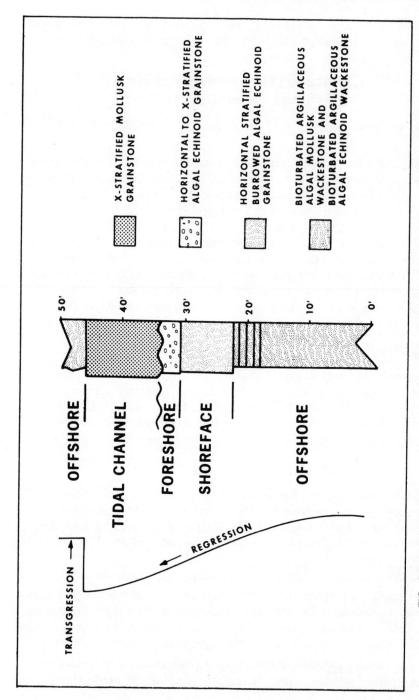

FIG. 3-2: Lower Cretaceous carbonate regressive shoreline sequence south of Sweetwater, Texas.

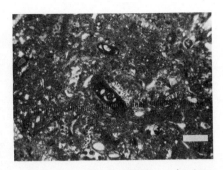

FIG. 3-3: Photomicrograph of bioturbated, argillaceous, algal-mollusk wackestone (offshore).

FIG. 3-4: Photomicrograph of bioturbated, argillaceous, algal-mollusk wackestone (offshore).

(White bar in the lower right corner of photomicrographs represents 1 mm)

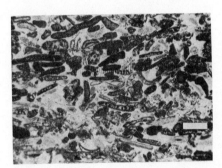
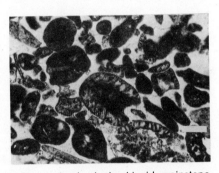

FIG. 3-5: Photomicrograph of well-sorted, medium-grained, algal-echinoid grainstone (shoreface).

FIG. 3-6: Photomicrograph of well-sorted, coarse-grained, algal-echinoid grainstone (foreshore).

(White bar in the lower right corner of photomicrographs represents 1 mm)

echinoid plates and spines with expitaxial cement (Fig. 3-6). Additional allochems present are peloids, gastropods, and pelecypod fragments.

High angle cross-stratified, mollusk grainstone. — tan, poorly sorted, very coarse-grained (0.25 to 20.0 mm), rounded pelecypod fragments with drusy and mosaic cement (Fig. 3-7). Other allochems are peloids and intraclasts.

Bioburbated, algal-echinoid wackestone. — codiacean green algae and echinoid fragments in a medium-gray, argillaceous, micritic matrix (Fig. 3-8). Other skeletal debris include minor gastropod and pelecypod fragments.

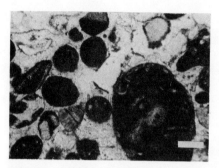 

FIG. 3-7: Photomicrograph of poorly sorted, very coarse-grained, mollusk grainstone (tidal channel deposit).

FIG. 3-8: Photomicrograph of bioturbated algal-echinoid wackestone (offshore).

(White bar in the lower right corner of photomicrographs represents 1 mm)

DEPOSITIONAL HISTORY

The bioturbated, argillaceous, algal-mollusk wackestones (Figs. 3-2, 3-3, and 3-4) represent offshore carbonate facies deposited below wave base. These offshore deposits are overlain by well-sorted, horizontally stratified, and burrowed (ophimorpha) algal-echinoid, shoreface grainstones (Figs. 3-2 and 3-5). Additional regression is indicated by the overlying foreshore, well-sorted, low angle cross-stratified, algal-echinoid grainstones (Figs. 3-2 and 3-6). The poorly sorted, high angle cross-stratified, mollusk grainstones (Figs. 3-2 and 3-7) are tidal channel deposits filling scours cut into the underlying foreshore deposits

The regressive carbonate shoreline sequence is overlain by offshore, bioturbated, argillaceous, algal-echinoid wackestones (Figs. 3-2 and 3-8) and indicates a transgressive cycle interrupting the underlying regressive carbonate sequence. This sequence is similar to modern (Reineck, and Singh, 1975, p. 316) and ancient (Howard, 1972, p. 215) regressive terrigenous clastic shoreline sequences.

Mississippian Upper Mission Canyon Shoreline Carbonates Truro Field Renville County, North Dakota

Twenty oil fields (Williston Basin Oil and Gas fields structural reference map, 1978) located in Renville County, North Dakota (northeastern Williston Basin) produce from the upper Mission Canyon (Mississippian) Formation. The upper Mission Canyon consists of a series of nearshore carbonate evaporite regressive cycles with the direction of regression to the west and southwest (Harris, et al., 1966, p. 2273). Traps are formed where porous, shoreface-foreshore grainstones and lagoonal dolomites pinch out up-dip into nonporous anhydritic tidal flat deposits. Accumulations of hydrocarbons are found where these traps exist across basinward-plunging anticlinal noses (Harris, 1966, et al., p. 2274–2275).

The three main upper Mission Canyon productive zones are, in ascending order: (1) Glenburn zone, (2) Sherwood zone, and (3) Bluell zone. These zones represent regressive paleoshorelines displaced basinward (seaward) through time (Wakefield, 1965, p. 139). Younger Mission Canyon production, therefore, occurs more basinward (Wakefield, 1965, p. 141).

The upper Mission Canyon Truro Field is located in sections 21 and 22 of T160N-R84W in Renville County, North Dakota. The field was discovered in 1976 and produces from the Bluell zone.

PETROGRAPHY

General The rocks of the Bluell zone in the Truro Field can be subdivided into five rock types: (1) oolitic-peloid grainstone, (2) crinoidal-peloid grainstone, (3) fossiliferous, anhydritic, pisolitic dolomite, (4) dolomitic anhydrite, and (5) dolomitic, anhydritic shale. The grainstones are subtidal, open-marine, shoreface-foreshore deposits, and the anhydritic dolomites represent lagoonal deposits. These shoreface-foreshore grainstones and lagoonal dolomites pinch-out to the northeast into nonporous, tidal flat, dolomitic anhydrite and dolomitic, anhydritic shales.

Detailed Description *Dolomitic anhydrite.*—white fibrous to bladed, fine-grained (5-10μ) crystals of anhydrite with minor stringers of brown microcrystalline dolomite (Fig. 3-9).

Dolomitic, anhydritic shale. — dark-gray, dolomitic shale with small nodules of white, fine-grained (5-10μ) anhydrite. Anhydrite also is present as white stringers of fine-grained anhydrite (Fig. 3-10).

Anhydritic, fossiliferous, pisolitic dolomite. — tan, microcrystalline, dolomite mudstone with pisolites and minor ostracods and gastropods. Both vadose (Fig. 3-11) and algal (Fig. 3-12) pisolites are present. The vuggy porosity has been partially filled with anhydrite and calcite. These pisolitic carbonates are the pay zone in the Truro Field.

Oolitic-peloid grainstone. — tan, well-sorted and rounded peloids, and minor oolites with intergranular porosity (Fig. 3-13). The oolites are

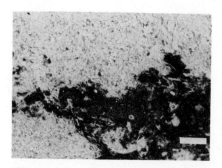
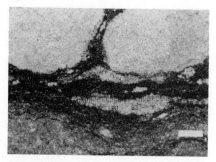

FIG. 3-9:   Photomicrograph of dolomitic anhydrite (tidal flat).
FIG. 3-10:   Photomicrograph of dolomitic anhydritic shale (tidal flat).
(White bar in the lower right corner of photomicrographs represents 1 mm)

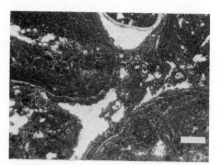

FIG. 3-11:   Photomicrograph of pisolitic, anhydritic dolomite with vuggy porosity (lagoonal). The pisolites present in this rock are vadose pisolites.
FIG. 3-12:   Photomicrograph of fossiliferous, anhydritic, pisolitic dolomite with vuggy porosity (lagoonal). The pisolites present in this rock are algal pisolites.
(White bar in the lower right corner of photomicrographs represents 1 mm)

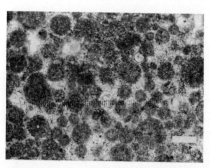
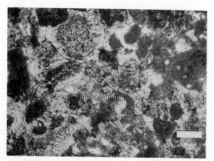

FIG. 3-13:   Photomicrograph of oolitic-peloid grainstone with intergranular porosity (foreshore).
FIG. 3-14:   Photomicrograph of crinoidal-peloid grainstone with intergranular porosity (shoreface-foreshore).
(White bar in the lower right corner of photomicrographs represents 1 mm)

micritized so that the concentric structure is faint. The very-fine, granular, calcite cement probably represents meniscus cement (see: Chapter 2).

Crinoidal-peloid grainstone. — tan, well-sorted peloids with crinoid debris partially cemented with epitaxial sparry calcite (Fig. 3-14). Additional allochems include bryozoan and brachiopod fragments and minor oolites. Minor amounts of micrite matrix are also present. Due to only partial cementation, some of the original intergranular porosity is preserved.

DEPOSITIONAL HISTORY

Figure 3-15 is a cross-section across the Truro Field. The sonic log in Figure 3-15 illustrates a rapid transition from porous carbonate grainstones and dolomites to nonporous anhydrites between the Wise Oil Co. Marie Solar No. 1 (Sec. 22 160N-84W) and the Colorado Oil Co. Solar A-1 (Sec. 22 160N-84W) wells. This transition is a paleoshoreline along which subtidal, open-marine, carbonate grainstones and lagoonal dolomites grade into tidal flat anhydrites. The transition, therefore, forms the up-dip seal on the reservoir (Fig. 3-15).

Figure 3-16 is a slice (paleogeographic) map at a 30-ft. slice interval below the State A Marker (Fig. 3-15). The slice interval represents the level of the pay zone in the Truro Field. The paleoshoreline delineated by this slice interval forms a trap as it crosses anticlinal noses (Fig. 3-16). Examination of Figure 3-15 reveals that a 10-foot slice map would displace the paleoshoreline to the southwest (marineward). The depositional sequence is regressive, and the paleoshoreline illustrated in Figures 3-15 and 3-16 developed during a period of stabilized sea level. The shoreline regression to the southwest is important since nonporous, tidal flat anhydrites were placed over the porous, carbonate grainstones and dolomites forming a seal (Fig. 3-15). During this regression, some of the subtidal lagoonal mudstones were exposed to vadose diagenesis as indicated by the presence of vadose pisolites (Fig. 3-11). Vadose pisolites have also been noted in the Mission Canyon Formation in the Glenburn Field in Renville and Bottineau counties of North Dakota (Gerhard, et al., 1978, p. 177–188).

The combination structural and stratigraphic trap outlined for the Truro Field is also present in the Mouse River Park (T162N - R86W and T162N - R85W) Field in Renville County, North Dakota, and in many other Mission Canyon oil fields of the northeastern Williston Basin (Harris, et al., 1966).

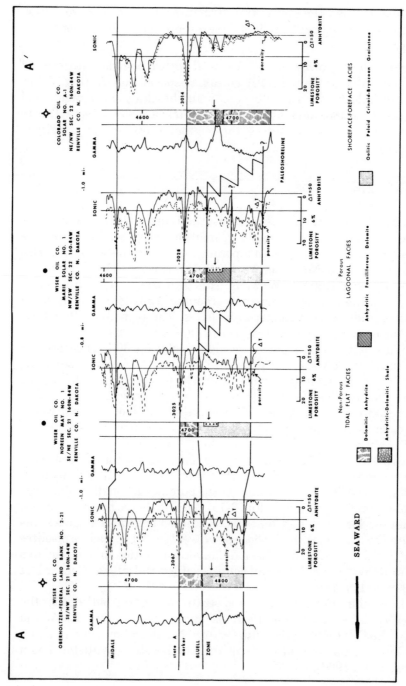

FIG. 3-15: Cross-section A-A' illustrating the facies change from porous subtidal and lagoonal carbonates to nonporous tidal flat anhydrites. Line of section illustrated in Figure 3-16. Arrows indicate 30-ft. slice interval mapped in Figure 3-16.

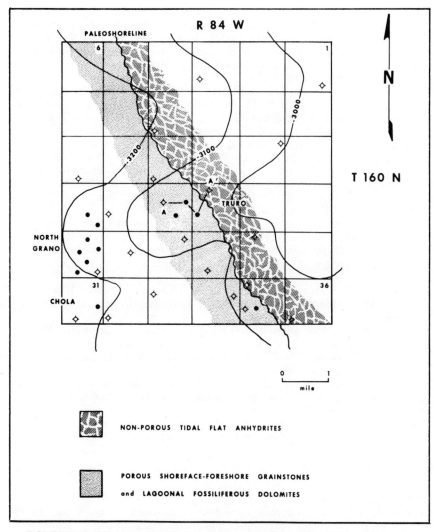

FIGURE 3-16: Paleogeographic map of the upper Mission Canyon Bluell zone at a slice interval 30 ft. below the State A marker (see: Fig. 3-15). Contours represent a structure map constructed on the State A marker (see: Fig. 3-15).

REFERENCES

Curray, J. R., 1969, Shorezone sand bodies, barriers, cheniers, and beach ridges *in* The new concepts of continental margin sedimentation: *Am. Geol. Institute Short Course Lecture Notes*, p. II-1 to II-19.

Evans, G., Murray, J. W., Biggs, N. E. J., Gate, R., and Bush, P. R., 1973, The
     oceanography, ecology, sedimentation, and geomorphology of parts of the
     Trucial Coast barrier island complex, Persian Gulf, p. 233–277 in Purser,
     B. H., ed., The Persian Gulf, Holocene carbonate sedimentation and
     diagenesis in a shallow epicontinental sea: Berlin, Heidelberg, and New
     York, Springer-Verlag, 471 p.
Gerhard, L. C., Anderson, S. B., and Berg. J., 1978, Mission Canyon porosity
     development, Glenburn Field, North Dakota Williston Basin: Williston
     Basin Symposium, Montana Geol. Soc., Billings Montana, p. 177–188.
Harris, S. H., Land, C. B., and McKeever, J. H., 1966, Relation of Mission
     Canyon stratigraphy to oil production in north-central North Dakota:
     Am. Assoc. Petroleum Geologists Bull., v. 50, no. 10, p. 2269–2276.
Howard, J. D., 1972, Trace fossils as criteria for recognizing shorelines in the
     stratigraphic record, p. 215–225 in Rigby, J. K. and Hamblin, W. K., eds.,
     Recognition of Ancient Sedimentary Environments: Soc. Econ. Paleon-
     tologists and Mineralogists, Spec. Pub. No. 16, 340 p.
Loreau, J. P., and Purser, B. H., 1973, Distribution and ultrastructure of
     Holocene ooids in the Persian Gulf, p. 279–328 in Purser, B. H., ed., The
     Persian Gulf, Holocene carbonate sedimentation and diagenesis in a
     shallow epicontinental sea: Berlin, Heidelberg, and New York, Springer-
     Verlag, 471 p.
Montana and Williston Basin Oil and Gas fields and Structural Reference
     Map, 1978: Montana Oil Journal and Technical Graphic Arts, Billings,
     Montana.
Moore, C. H. Jr., 1969, Depositional Environment and Depositional History
     Lower Cretaceous Shallow Shelf Carbonate Sequence West-Central
     Texas: Guidebook Dallas Geological Soc., Am. Assoc. Petroleum
     Geologists meeting, 135 p.
Reineck, H. E., and Singh, I. B., 1975, The Coast, p. 280–305 in Depositional
     Sedimentary Environments: Berlin, Heidelberg, and New York,
     Springer-Verlag, 439 p.
Wakefield, W. W., 1965, Much good oil-hunting land left in North Dakota's
     Mississippian: Oil and Gas Jour., v. 63, no. 11, p. 138–146.

SUGGESTED READING

Shoreline Deposits
Curray, J. R., 1969, Shorezone sand bodies, barriers, cheniers, and beach
     ridges, in The New Concepts of Continental Margin Sedimentation: Am.
     Geol. Institute Short Course Lecture Notes, p. II-1 to II-19.
Evans, G., Murray, J. W., Biggs, N. E. J., Gate, R., and Bush, P. R., 1973, The
     oceanography, ecology, sedimentation, and geomorphology of parts of the
     Trucial Coast barrier island complex, Persian Gulf, p. 233–277 in Purser,
     B. H., ed. The Persian Gulf, Holocene carbonate sedimentation and
     diagenesis in a shallow epicontinental sea; Berlin, Heidelberg, and New
     York, Springer-Verlag, 471 p.
Moore, C. H., Jr., 1969 Depositional Environment and Depositional History
     Lower Cretaceous Shallow Shelf Carbonate Sequence West-Central
     Texas: Guidebook Dallas Geological Soc., Am. Assoc. of Petroleum
     Geologists meeting, 135 p.

4

Carbonate Tidal
Flat Model

General

A carbonate tidal flat is a marshy land area which is covered and uncovered by the rise and fall of the tide. Tidal flats develop along shorelines protected from waves (i.e. bays and lagoons) and on shallow carbonate shelves where wave energy is low. A recent sediment model of protected coastal tidal flats is the SE Persian Gulf (Illing, et al., 1965; Shinn, 1973), while a recent model for tidal flats on a shallow carbonate shelf (Fig. 4-4) is the Bahama Platform (Shinn, et al., 1965; Shinn, et al., 1969).

Carbonate sediments accumulate in three major environments associated with tidal flats: (1) subtidal — in open-marine or lagoons permanently below low tide, (2) intertidal—between normal high and low tides, and (3) supratidal—above normal high tide, but within the range of spring and storm tides. The supratidal environment is commonly referred to as a sabkha.

Modern subtidal sediment in more protected areas consists of soft pelletal mud and silt-size carbonate and contains varying amounts of skeletal grains (Shinn, 1973, p. 181). The sediments are often gray in color and smell of H₂S. They lack primary sedimentary structures and are highly burrowed (Shinn, 1973, p. 181).

Shinn (1973), p. 183) has noted that modern intertidal sediments resemble subtidal sediments because of their pelletal mud and carbonate silt composition; they normally contain less skeletal debris than subtidal sediments and the fauna is more restricted. The sediments are light-tan (oxidized) and have iron stained root burrows, but with fewer animal burrows than subtidal sediments. "Birdseye" vugs (fenestrate fabric; Fig. 4-5) are often present; laminations (algal), except in the upper parts of the intertidal zone, are uncommon.

Modern supratidal sediments, although predominantly mud, are highly variable. Both skeletal debris, except for displaced faunas, and burrowing are rare. The sediments are characterized by: laminations, birdseye vugs, mud cracks, and a light-tan color (Shinn, 1973, p. 183). Syngenetic dolomite and evaporites (gypsum and anhydrite) are common in supratidal sediments.

The presence or absence of evaporites in the supratidal environment is controlled by climate since sabkhas in the arid Persian Gulf have evaporites (Purser and Evans, 1973, p. 225) and those in the more humid Bahamas do not (Shinn, et al., 1965). As the tidal flats prograde seaward, they leave a characteristic regressive "sabkha sequence" composed of subtidal, intertidal, and supratidal sediments (Purser and Evans, 1973, p. 231).

Porosity can be developed in all three environments and is usually secondary porosity due to dolomitization or leaching. It is the author's experience, however, that better porosity is more often developed in the subtidal facies (Asquith, et al., 1978, p. 72). This occurs because of two factors: (1) the dolomites in the supratidal and intertidal facies are normally microcrystalline, while in the subtidal facies, the coarser, sucrosic dolomites often develop; and (2) the supratidal and intertidal facies frequently have their porosity plugged by secondary anhydrite cement.

The carbonate tidal flat environment is a common reservoir for oil and gas and, consequently, is an important carbonate model for the petroleum geologist. The following is a partial list of some of the better known tidal flat reservoirs in North America: (1) Ordovician Stony Mountain and Silurian Interlake formations, Williston Basin (Roehl, 1967), (2) Permian upper Clear Fork Formation, Permian Basin, Texas (Lucia, 1972), (3) Ordovician Red River Formation, Williston Basin, Montana (Asquith, et al., 1978), (4) Jurassic Buckner Formation, Gulf Coast (Dickinson, 1968), (5) Mississippian Charles Formation, Williston Basin (Selley, 1970, p. 139–140), and (6) Devonian Duperow Formation, Williston Basin (Wilson, 1975, p. 298).

Ordovician Red River C and D Zones Big Muddy Creek Field, Roosevelt County, Montana

INTRODUCTION

The Big Muddy Creek Field, located in the northwest part of the Williston Basin (T29N-R55E) in Roosevelt County, Montana (Fig. 4-1), is developed on two isolated seismic highs occurring along a northwest trending anticlinal ridge (Fig. 4-2). Oil production is from surcrosic dolomite reservoirs developed in the Ordovician Red River C and D zones below the Red River C anhydrite. The Red River Formation is underlain by the Ordovician Winnipeg Formation and is overlain by the Ordovician Stony Mountain Shale.

PETROGRAPHY

General The carbonate rocks of the Ordovician Red River C and D zones can be subdivided into four different rock types: (1) anhydritic, laminated dolomite; (2) dolomitic mudstone; (3) fossiliferous, sucrosic dolomite; and (4) mottled, dolomitic, crinoid-brachiopod packstone-wackestone. The anhydritic, laminated dolomite was deposited in a supratidal environment; the dolomitic mudstone represents intertidal deposition; the fossiliferous, sucrosic dolomite and mottled, dolomitic, crinoid-brachiopod packstone-wackestone indicates a subtidal environment. These rock types were deposited in a regressive "sabkha sequence" as illustrated in the Anadarko Production Alpar-State No. 1-28 (Fig. 4-3).

Detailed Descriptions *Anhydritic, laminated dolomite.* — microcrystalline dolomite ($<10\mu$) with fine, distinct to faint laminations that sometimes display minor brecciation (Fig. 4-6); the laminations probably are algal in origin. Anhydrite is present as fine-grained (10-20μ) fibrous and bladed crystals and is commonly interlaminated with the microcrystalline dolomite. Thin zones (2-4 ft.) appearing in the anhydritic, laminated dolomite facies have greater than 50 percent anhydrite with anhydrite nodules frequently present. These zones are evident on the log cross-section (Fig. 4-10) and exhibit the typical CNL*-FDC* log response for anhydrite (i.e. CNL*=0 and FDC* < -10). This lithofacies represents a supratidal environment (Asquith, et al., 1978).

*A Mark of Schlumberger

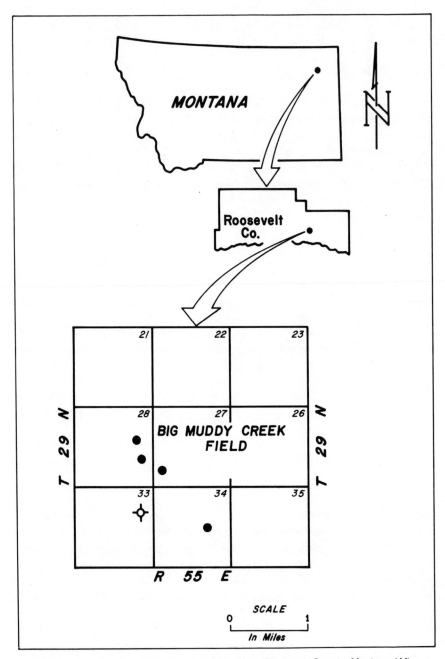

FIG. 4-1: Index map of Big Muddy Creek Field, Roosevelt County, Montana (After: Asquith, et al., 1978).

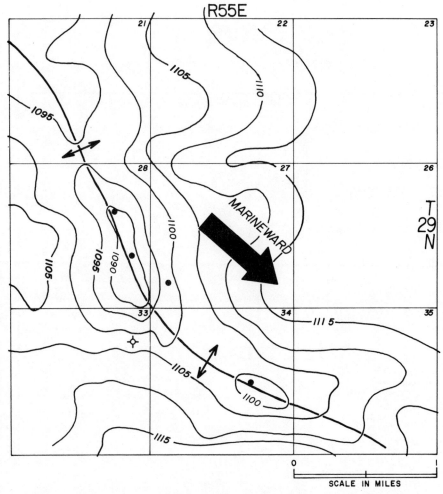

FIG. 4-2: Greenhorn to Winnipeg isochron map of Big Muddy Creek structure (contour interval 5 milliseconds). (after: Asquith, et al., 1978).

Dolomitic mudstone. — micrite matrix partly to completely replaced by microcrystalline dolomite (10-20μ). The dolomitic mudstones (Fig. 4-7) contain minor (<10 percent) pelecypods, ostracods, peloids, and gastropods. Large (up to 1000μ), bladed crystals of anhydrite often occur scattered within the micrite matrix. The dolomitic mudstones were deposited in an intertidal environment (Asquith, et al., 1978).

Fossiliferous, sucrosic dolomite. — crinoid and brachiopod debris in a micrite matrix replaced by finely crystalline (40-60μ), sucrosic dolo-

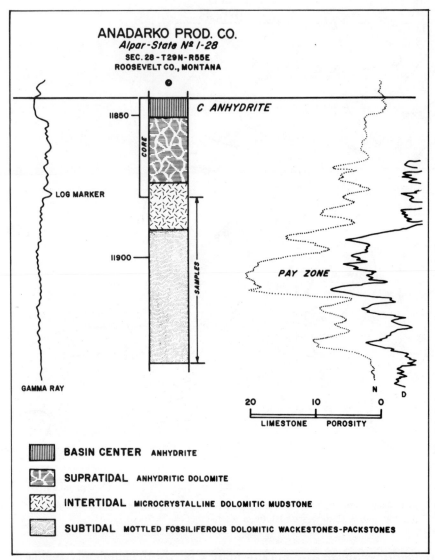

FIG. 4-3: Carbonate rock types and corresponding log responses for the Red River C and D zones in the Anadarko Alpar-State 1-28. The C zone is located above the log marker and the D zone below. (after: Asquith, et al., 1978).
N = neutron   D = density

mite (Fig. 4-8). Minor pelecypod and ostracod fragments also occur. The fossil debris is often leached forming fossilmoldic porosity. The fossiliferous, sucrosic dolomites represent subtidal facies and are the

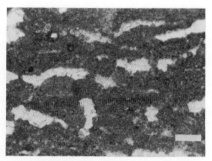

FIG. 4-4: Photograph of isolated supratidal flats on the Bahama Platform.
FIG. 4-5: Photomicrograph of intertidal fenestrate fabric from the Permian Tansill Formation in the Gifford, Mitchell, and Wisenbaker No. 1 Little Wolf, Winkler County, Texas (depth 3,184 ft.).

(White bar in the lower right corner of photomicrographs represents 1 mm)

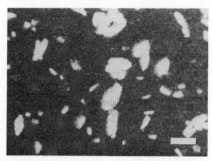

FIG. 4-6: Photomicrograph of supratidal, anhydritic, laminated dolomite in Red River C-zone (11,862 ft.) Anadarko Production Company Alpar State No. 1-28, Sec. 28, T29N-R55E, Roosevelt County, Montana. The upper part of the slide contains a large anhydrite nodule.
FIG. 4-7: Photomicrograph of intertidal dolomitic mudstone with porphyroblastic anhydrite in Red River C-zone (11,880 ft.) Anadarko Production Company Alpar State No. 1-28, Sec. 28, T29N-R55E, Roosevelt County, Montana.

(White bar in the lower right corner of photomicrographs represents 1 mm)

reservoir rocks in the southeast portion of the Big Muddy Creek Field in the Alpar Resources Waldow No. 1 (Asquith, et al., 1978).

Mottled, dolomitic, crinoid-brachiopod packstone-wackestone. — crinoid and brachiopod fragments in a slightly argillaceous, micrite matrix (Fig. 4-9). Also, contains minor amounts of ostracods, bryozoa, trilobites, pelecypods, and dasycladacean algae debris. The mottled texture is due to horizontal and vertical burrows (10-20mm in diameter) that have been selectively replaced by finely sucrosic (40-60μ) dolomite. The mottled, dolomitic, crinoid-brachiopod packstone-wackestones are subtidal carbonates and, when the micrite matrix is

 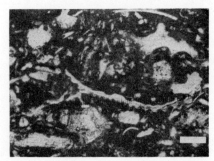

FIG. 4-8: Photomicrograph of subtidal, fossiliferous, sucrosic dolomite in Red River C-zone (12,005) Alpar Resources Waldow No. 1, Sec. 34, T29N-R55E, Roosevelt County, Montana.
FIG. 4-9: Photomicrograph of subtidal, mottled, dolomitic, crinoid-brachiopod wackestone-packstone in Red River D-zone (12,052 ft.) Alpar Resources Waldow No. 1, Sec. 34, T29N-R55E, Roosevelt County, Montana.

(White bar in the lower right corner of photomicrographs represents 1 mm)

extensively dolomitized, contain the reservoir rocks on the main Big Muddy Creek structure (Asquith et al., 1978).

DEPOSITIONAL HISTORY

In order to understand the changing stratigraphic position of the reservoirs in the Big Muddy Creek Field, it is important to know that, regionally, the marineward direction in the area is southeast (Fig. 4-2). Regressive tidal flat sequences, like the one developed in the Big Muddy Creek Field, prograde marineward through time whenever the rate of sediment accumulation exceeds the ability of the marine processes (i.e. waves and tides) to redistribute the sediment. Progradation can take place during times of either stable or falling sea level. This marineward progradation is illustrated by the strati-graphically higher position of the pay zone (Fig. 4-10) in the Anadarko Production Barry Rogney No. 1-A relative to the Anadarko Produc-tion Alpar-State No. 1-28. Pay zones in the Field are developed along the contact between the subtidal and intertidal facies (Fig. 4-10), because the facies prograde seaward over the structure. The location of the pay zones along this contact may be the result of the seaward movement of dolomitizing magnesium-rich meteoric waters from the supratidal facies (Asquith, et al., 1978, p. 73). This model is referred to as "hydrodynamic dolomitization," (Jacka and Franco, 1975; Barone, 1976).

The reservoir in the Alpar Resources Waldow No. 1, located in the southeast portion of the field, is in a transgressive marine tongue

(Asquith, et al., 1978, p. 73) stratigraphically higher than the pay zones on the main Big Muddy Creek structure (Fig. 4-10). Transgressive marine tongues such as this have been referred to as "kick backs" by Irwin (1965, p. 456), because they represent a minor transgressive event in the overall regressive cycle.

Irwin (1965, p. 451) and Purser (1973, p. 157–177) have observed in both recent and ancient carbonates that topographic highs can greatly affect carbonate sedimentation by creating local shallow water conditions. Such topographic highs influence carbonate facies, carbonate rock type, and diagenesis. Purser (1973, p. 177) noted that topographic highs in the Persian Gulf can be localized areas of dolomitization and vadose leaching. These observations by Purser and Irwin may be the reason why there is often a greater degree of dolomitization and better reservoir development on the crest of Red River structures (Ballard, 1969, p. 23) such as Big Muddy Creek.

The facies pattern developed on the Big Muddy Creek structure during deposition of the Red River C zone demonstrates the relationship between topographic highs and carbonate facies (Fig. 4-11). Examination of a 10-ft. slice interval (Fig. 4-11) below the Red River C anhydrite (level of pay zone in the Alpar Resources Waldow No. 1) reveals the affect of structure on the distribution of carbonate facies.

The larger and higher of the two structures in the Big Muddy Creek Field has supratidal and intertidal facies developed, while off this structure to the southeast, subtidal facies are present (Fig. 4-11). This change in facies to the southeast is illustrated in Figure 4-10 where there is a progressive change in environments from supratidal to intertidal to subtidal moving from the Anadarko Production Alpar-State No. 1-28 southeast to the Alpar Resources Waldow No. 1 along a 10-ft. slice interval. An increase in CNL*-FDC* porosity (Fig. 4-10) accompanies this facies change.

The decision to drill the Alpar Resources Waldow No. 1. was based on development of the CNL*-FDC* porosity and the presence of subtidal fossiliferous sucrosic dolomite in the C zone of the Anadarko Krueger A-1, plus the presence of a down-dip isolated seismic closure (Fig. 4-11; Asquith, et al., 1978). Examination of Figures 4-10 and 4-11 indicates an up-dip pinch out of porosity to the northwest in the Red River C zone. Thus, the trap in the Alpar Resources Waldow No. 1 is at least partly stratigraphic.

The depositional history of the Red River C and D zones is illustrated by a series of slice maps (Asquith, et al., 1978) at 5 to 50-ft.

*A Mark of Schlumberger

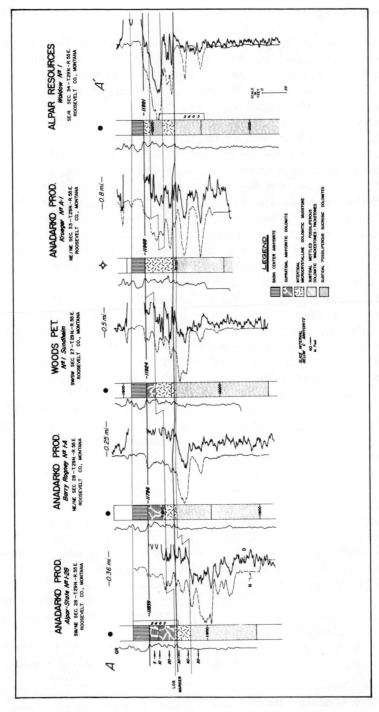

FIG. 4-10: Cross-section A-A' along the axis of the Big Muddy Creek structure. The line of cross-section is shown in Figure 4-11 (after: Asquith, et al., 1978).

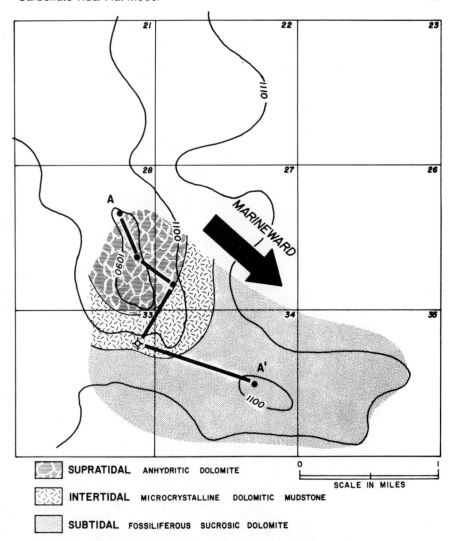

FIG. 4-11: Distribution of Red River depositional environments at a slice interval 10 ft. below the C anhydrite. Contours are Greenhorn to Winnipeg isochrons. (Contour interval 10 milliseconds). A-A′ line of cross-section is illustrated in Figure 4-10 (after: Asquith, et al., 1978).

intervals below the Red River C anhydrite (see: cross-section A-A′, Fig. 4-10). At time one (i.e. 50-ft. slice interval representing the pay zone level in the northern-most well of the field — Anadarko Production Alpar-State No. 1-28), subtidal, mottled, dolomitic, crinoid-brachiopod wackestone-packstone covered the entire Big Muddy Creek structure (Fig. 4-12). At time two (40-ft. slice representing the level of the pay zone in the Anadarko Production Barry Rogney No. 1-A and the Woods Petroleum No. 1 Sundheim), intertidal facies appear in the northwest part of the structure (Fig. 4-13).

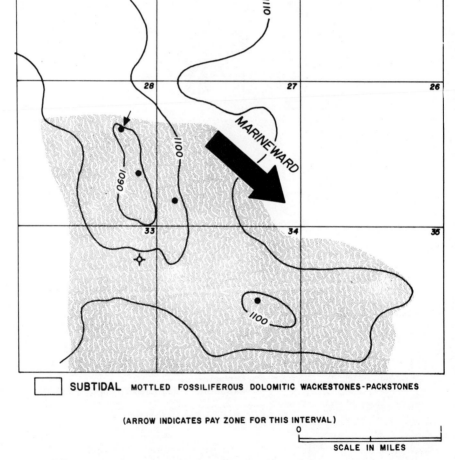

SUBTIDAL MOTTLED FOSSILIFEROUS DOLOMITIC WACKESTONES-PACKSTONES

(ARROW INDICATES PAY ZONE FOR THIS INTERVAL)

SCALE IN MILES

FIG. 4-12: Distribution of Red River depositional environments at a slice interval 50 ft. below the C anhydrite. Contours are Greenhorn to Winnipeg isochrons. (Contour interval 10 milliseconds). (after: Asquith, et al., 1978).

These intertidal facies represent the start of the southeast progradation of the regressive tidal flat cycle. At time three (30-ft. slice), nonporous, intertidal, dolomitic mudstones covered the entire structure (Fig. 4-14). These nonporous, intertidal, dolomitic mudstones

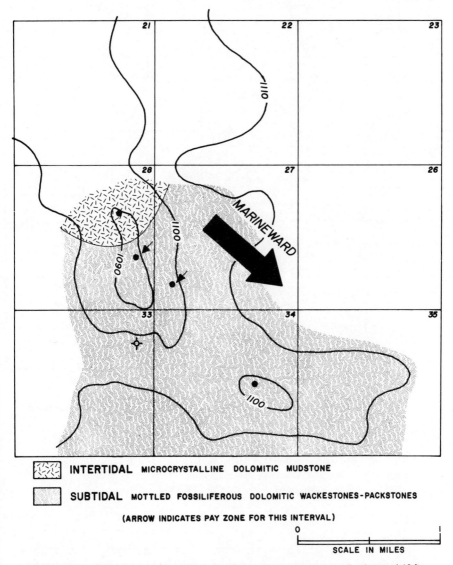

INTERTIDAL MICROCRYSTALLINE DOLOMITIC MUDSTONE

SUBTIDAL MOTTLED FOSSILIFEROUS DOLOMITIC WACKESTONES-PACKSTONES

(ARROW INDICATES PAY ZONE FOR THIS INTERVAL)

SCALE IN MILES

FIG. 4-13: Distribution of Red River depositional environments at a slice interval 40 ft. below the C anhydrite. Contours are Greenhorn to Winnipeg isochrons. (Contour interval 10 milliseconds). (After: Asquith, et al., 1978).

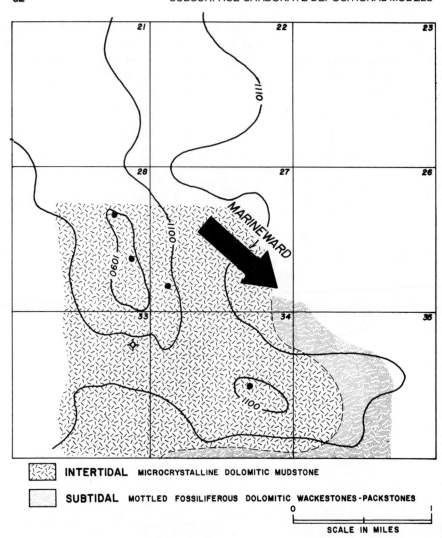

INTERTIDAL MICROCRYSTALLINE DOLOMITIC MUDSTONE

SUBTIDAL MOTTLED FOSSILIFEROUS DOLOMITIC WACKESTONES-PACKSTONES

SCALE IN MILES

FIG. 4-14: Distribution of Red River depositional environments at a slice interval 30 ft. below the C anhydrite. Contours are Greenhorn to Winnipeg isochrons. (Contour interval 10 milliseconds). (after: Asquith, et al., 1978).

form the seal on the D zone reservoirs in the main Big Muddy Creek structure. At times four and five (20 and 10-ft. slice intervals), there is a minor marine transgression within the overall regressive cycle that is present only in the southeast portion of the field (Figs. 4-15

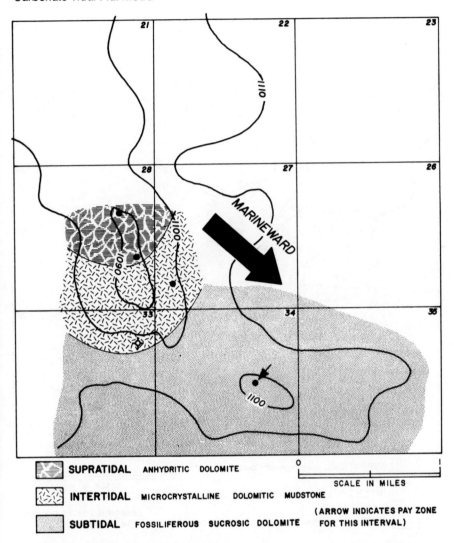

SUPRATIDAL ANHYDRITIC DOLOMITE

INTERTIDAL MICROCRYSTALLINE DOLOMITIC MUDSTONE

SUBTIDAL FOSSILIFEROUS SUCROSIC DOLOMITE

0   1
SCALE IN MILES

(ARROW INDICATES PAY ZONE FOR THIS INTERVAL)

FIG. 4-15: Distribution of Red River depositional environments at a slice interval 20 ft. below the C anhydrite. Contours are Greenhorn to Winnipeg isochrons. (Contour interval 10 milliseconds). (after: Asquith, et al., 1978).

and 4-16). This transgressive tongue contains the C zone reservoir in the Alpar Resources Waldow No. 1.

Time six (5-ft. slice) illustrates the final regressive cycle for the Red River C zone when supratidal and intertidal facies covered the

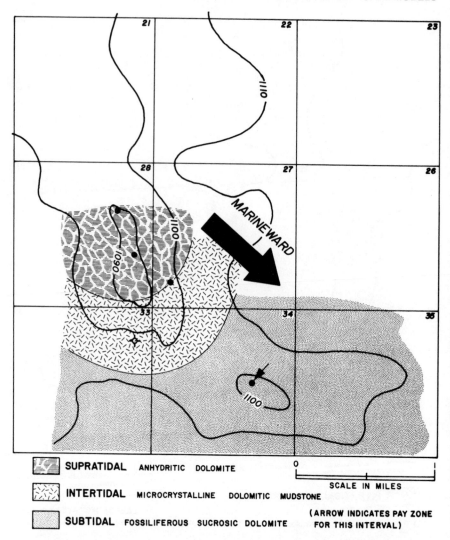

FIG. 4-16: Distribution of Red River depositional environments at a slice interval 10 ft. below the C anhydrite. Contours are Greenhorn to Winnipeg isochrons (Contour interval 10 milliseconds). (after: Asquith, et al., 1978).

entire structure (Fig. 4-17). These nonporous, anhydritic, laminated dolomites and dolomitic mudstones form the seal on the C zone reservoir in the Alpar Resources Waldow No. 1.

Asquith, et al. (1978) suggest that there are two stages of dolomitization in the C and D zones in the Big Muddy Creek Field.

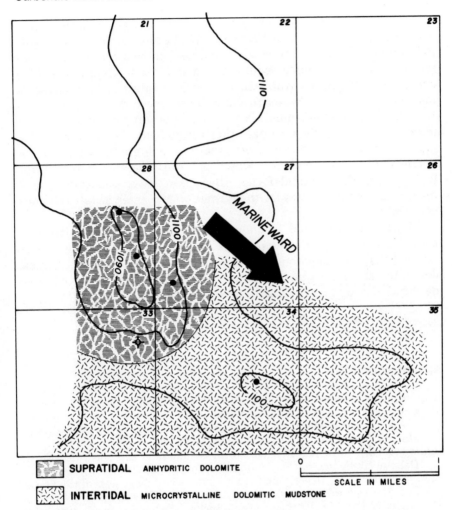

FIG. 4-17: Distribution of Red River depositional environments at slice interval 5 ft. below the C anhydrite. Contours are Greenhorn to Winnipeg isochrons (Contour interval 10 milliseconds). (after: Asquith, et al., 1978).

The syngenetic supratidal dolomite may be the result of evaporation that precipitated calcium sulfate ($CaSO_4$) and increased the magnesium/calcium ratio from 5.0 (normal sea water) to over 10.0. When this happens, microcrystalline dolomite begins to replace the calcium carbonate sediments (Illing, et al., 1965).

The intertidal and subtidal dolomites developed from diagenetic replacement by magnesium-rich meteoric water ("hydrodynamic dolomitization," Jacka and Franco, 1975; Barone, 1976). In the hydrodynamic dolomitization model, magnesium-rich meteoric waters migrate seaward through sediment at a shallow depth of burial. The enrichment of ground water in magnesium takes place because of movement of meteoric water through the magnesium-rich sabkha.

The intertidal sediments are replaced by microcrystalline dolomite because of multiple nucleation throughout the sediment. The multiple nucleation occurs when there is a high concentration of magnesium in the meteoric water. As the meteoric water moves further into the subtidal facies, there is a lower concentration of magnesium which allows for the formation of coarser, isolated rhombs of dolomite.

Both the supratidal syngenetic microcrystalline dolomites and the intertidal diagenetic microcrystalline dolomites in the Big Muddy Creek Field are nonporous, but the coarser grained subtidal sucrosic dolomites are porous and so form the reservoir rock. For a detailed description of Red River diagenesis see: Kendall (1976, p. 57–59).

REFERENCES

Asquith, G. B., Parker, R. L., Gibson, C. R., and Root, J. R., 1978, Depositional history of the Ordovician Red River C and D zones Big Muddy Creek Field Roosevelt County, Montana: *Williston Basin Symposium, Montana Geol. Soc.*, Billings, Montana, p. 71–76.

Ballard, W. W., 1969, Red River of northeast Montana and northwest North Dakota: *Montana Geol. Soc., 20th Ann. Conf.*, p. 15–24.

Barone, W. E., 1976, Depositional environments and diagenesis of the Lower San Andres Formation: *M.S. thesis*, Texas Tech Univ., 93 p.

Dickinson, K. R., 1968, Petrology of the Buckner Formation in adjacent parts of Texas, Louisiana, and Arkansas: *Jour. Sedimentary Petrology*, v. 38, p. 555–567.

Illing, L. V., Wells, A. J., and Taylor, J. C. M., 1965, Penecontemporary dolomite in the Persian Gulf, p. 89–111 *in* Pray, L. C. and Murray, R. C., eds., *Dolomitization and Limestone Diagenesis, A symposium:* Soc. of Econ. Paleontologists and Mineralogists, Spec. Pub. No. 13, 180 p.

Irwin, M. L., 1965, General theory of epeiric clear water sedimentation: *Am. Assoc. Petroleum Geologists, Bull.*, v. 49, no. 4, p. 445–459.

Jacka, A. D. and Franco, L. A., 1975, Deposition and diagenesis of Permian evaporites and associated carbonates and clastics on shelf areas of the Permian Basin *in* 4th Symposium on Salt: *Northern Ohio Geological Society*, Cleveland, Ohio, p. 67–89.

Kendall, A. C., 1976, The Ordovician carbonate succession (Big Horn Group) of southeastern Saskatchewan: Dept. of Mineral Resources, *Saskatchewan Geological Survey*, Rep. No. 180, 185 p.
Lucia, F. J., 1972, Recognition of evaporite-carbonate shoreline sedimentation, p. 160–191 *in* Rigby, J. K., and Hamblin, W. K., *eds., Recognition of Ancient Sedimentary Environments:* Soc. of Econ. Paleontologists and Mineralogists, Spec. Pub. No. 16, 340 p.
Purser, B. H., 1973, Sedimentation around bathymetric highs in the southern Persian Gulf, p. 157–178 *in* Purser, B. H., *ed., The Persian Gulf, Holocene carbonate sedimentation and diagenesis in a shallow epicontinental sea:* Berlin, Heidelberg, and New York, Springer-Verlag, 471 p.
―――――――――――, and Evans, G., 1973, Regional sedimentation along the Trucial Coast, SE Persian Gulf, p. 211–232 *in* Purser, B. H., *ed., The Persian Gulf, Holocene carbonate sedimentation in a shallow epicontinental sea:* Berlin, Heidelberg, and New York, Springer-Verlag, 471 p.
Roehl, P. O., 1967, Stony Mountain (Ordovician) and Interlake (Silurian) facies analogs of recent low-energy marine and subaerial carbonates, Bahamas: *Am. Assoc. Petroleum Geologists, Bull.*, v. 51, p. 1979–2032.
Selley, R. C., 1970, Carbonate shoreline and shelf deposits *in Ancient Sedimentary Environments:* London, Chapman and Hall Ltd., 237 p.
Shinn, E. A., Ginsburg, R. N., and Lloyd, R. M., 1965, Recent supratidal dolomite from Andros Island, Bahamas, p. 112–123 *in* Pray, L. C., and Murray, R. C., *eds., Dolomitization and Limestone Diagenesis, A symposium:* Soc. of Econ. Paleontologists and Mineralogists, Spec. Pub. No. 13, 180 p.
―――――――――――, Lloyd, R. M., and Ginsburg, R. N., 1969, Anatomy of a modern carbonate tidal flat: *Jour. Sedimentary Petrology*, v. 39, no. 3, p. 1202–1228.
―――――――――――, 1973, Carbonate coastal accretion in an area of longshore transport, NE Qatar, Persian Gulf, p. 179–192 *in* Purser, B. H., *ed., The Persian Gulf, Holocene carbonate sedimentation and diagenesis in a shallow epicontinental sea:* Berlin, Heidelberg, and New York, Springer-Verlag, 471 p.
Wilson, J. L., 1975, *Carbonate Facies in Geologic History:* Berlin, Heidelberg, and New York, Springer-Verlag, 471 p.

SUGGESTED READING

Tidal Flats
Ginsburg, R. N., *ed.*, 1975, *Tidal deposits, A casebook of recent examples and ancient counterparts:* Berlin, Heidelberg, New York, Springer-Verlag, 428 p.
Illing, L. V., Wells, A. J., and Taylor, J. C. M., 1965, Penecontemporary dolomite in the Persian Gulf, p. 89–111 *in* Pray, L. C., and Murray, R. C., *eds., Dolomitization and Limestone Diagenesis, A symposium:* Soc. of Econ. Paleontologists and Mineralogists, Spec. Pub. No. 13, 180 p.
Lucia, F. J., 1972, Recognition of evaporite-carbonate shoreline sedimentation, p. 160–191 *in* Rigby, J. K., and Hamblin, W. K., *eds., Recognition of*

Ancient Sedimentary Environments, A symposium: Soc. of Econ. Paleontologists and Mineralogists, Spec. Pub. No. 16, 340 p.

Roehl, P. O., 1967, Stony Mountain (Ordovician) and Interlake (Silurian) facies analogs of recent low-energy marine and subaerial carbonates, Bahamas: *Am. Assoc. Petroleum Geologists, Bull.*, v. 51, p. 1979–2032.

Shinn, E. A., Lloyd, R. M., and Ginsburg, R. N., 1969, Anatomy of a modern carbonate tidal flat: *Jour. Sedimentary Petrology*, v. 39, no. 3, p. 1202–1228.

5

Oolite Shoal Model

General

Shoals are submarine topographic highs detached from the shoreline and may or may not be exposed during low tide. Some shoals are composed of oolites, which are spherical to ellipsoidal bodies 0.25 to 2.0 mm in diameter, that may have a nucleus, and have concentric or radial structure or both (Glossary of Geology, 1962). This chapter is concerned with the type of oolites that form shoals in high energy environments in the presence of bi-directional tidal currents.

Recent models of this type of oolite are the Bahamian oolite shoals (Purdy, 1961; Hine, 1977) and the oolite bars (shoals) of the SE Persian Gulf that are associated with reversible current systems (Loreau and Purser, 1973, p. 317). In addition to oolite shoals, oolites also form in tidal channels associated with coastal barrier islands (Loreau, and Purser, 1973, p. 285; see: Chapter 3).

Loreau and Purser (1973, p. 317–318) suggested that the process of "oolitization" in the tidal environment is not constant, but takes place in stages, depending upon whether the particle is on an agitated bar crest or is within a lower energy depression. They proposed that aragonite precipitates in a more or less radial orientation on the oolite by "in situ" growth in the low energy interbar depressions. The radial aragonite compacts and reorients (tangentially) when the oolite is returned to the agitated crest of the oolite bars. The bi-directional tidal currents are necessary to the process because they retain the oolite in the "oolitizing" system of alternating low and high energy environments (Loreau and Purser, 1973, p. 318).

Oolitic carbonates are important reservoir rocks for many oil and gas fields in the United States. The oolitic Smackover Limestone (Jurassic) is the pay zone in the Magnolia Field in Columbia County, Arkansas. In the Carthage gas Field of Panola County, Texas, one of the important pay formations is the upper Pettet (Lower Cretaceous) oolitic limestone.

The McCloskey Formation (Upper Mississippian) is a widespread and productive oolitic limestone in the Illinois Basin (Leverson, 1961, p. 116). The hydrocarbons in these fields are trapped in the pores between oolites (intergranular porosity) or in the casts left by dissolved oolites (oomoldic porosity) or a combination of both (Leverson, 1961, p. 116).

Lower Permian Council Grove B-Zone Oolite Shoal, Ochiltree County, Texas

INTRODUCTION

The Council Grove B-Zone oolite shoal is located in the northwest corner of the Anadarko Basin in Ochiltree County of the Texas Panhandle along the Oklahoma-Texas border. The Council Grove B-zone consists of an oolitic limestone within a sequence of Lower Permian (Wolfcamp) limestones and calcareous shales.

PETROGRAPHY

General The Lower Permian Council Grove B-zone, and the rocks above and below it, can be subdivided into four carbonate rock types: (1) argillaceous, crinoid-bryozoan wackestone; (2) oolitic-crinoid-bryozoan wackestone; (3) oolite grainstone; and (4) argillaceous, algal-crinoid-bryozoan wackestone. Figures 5-1 and 5-2 illustrate the vertical sequence of the different carbonate rock types and the associated sedimentary structures from the Crest Exploration Company Caldwell 2-1100.

Detailed Descriptions *Argillaceous, crinoid-bryozoan wackestone.* — dark-gray, bioturbated wackestones containing crinoid and bryozoan fragments scattered in an argillaceous, micrite matrix (Fig. 5-3). Also contains minor echinoid spines, fusulinids, and brachiopod fragments. This rock type underlies the Council Grove B-zone (Asquith, et al., 1977).

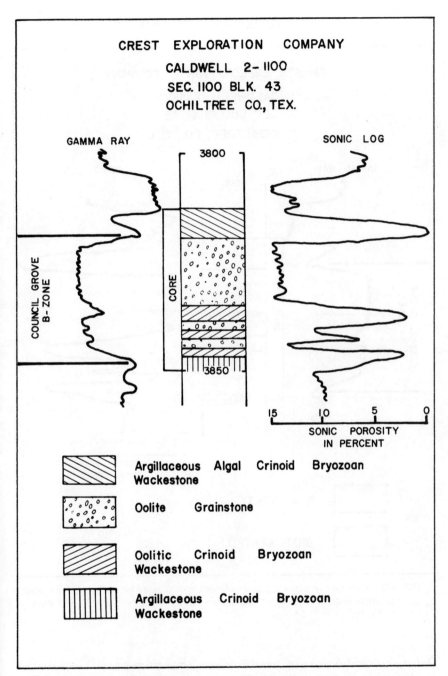

FIG. 5-1: Gamma ray, sonic log, and carbonate rock types for the Lower Permian Council Grove B-zone in the Crest Exploration Company Caldwell 2-1100, Ochiltree County, Texas (after: Asquith, et al., 1977).

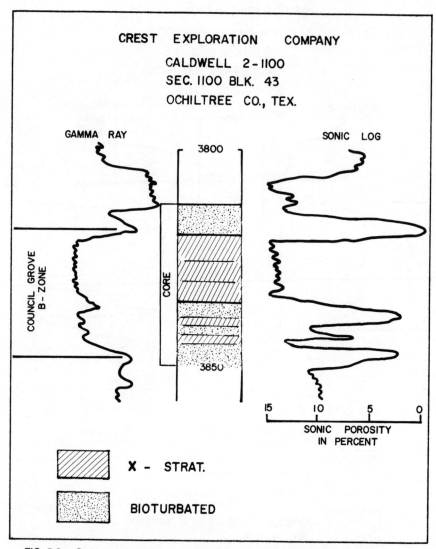

FIG. 5-2: Gamma ray, sonic log, and sedimentary structures for the Lower Permian Council Grove B-zone in the Crest Exploration Company Caldwell 2-1100, Ochiltree County, Texas.

Oolitic-crinoid-bryozoan wackestone. — medium-gray, bioturbated wackestones with micritized (Blatt, et al., 1972, p. 423) oolites, plus crinoid and bryozoan fragments in a micrite matrix (Fig. 5-4). Other fossil debris include minor brachiopod fragments, echinoid spines,

FIG. 5-3: Photomicrograph of bioturbated, argillaceous, crinoid-bryozoan wackestone Crest Exploration Company Caldwell 2-1100 (depth, 3849 ft.).

FIG. 5-4: Photomicrograph of bioturbated, oolitic-crinoid-bryozoan wackestone Crest Exploration Company Caldwell 2-1100 (depth, 3842 ft.).

(White bar in the lower right corner of photomicrographs represents 1 mm)

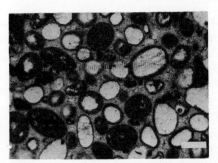 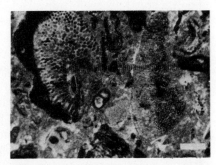

FIG. 5-5: Photomicrograph of cross-stratified oolite grainstone with oomoldic porosity Crest Exploration Company Caldwell 2-1100 (depth, 3828 ft.).

FIG. 5-6: Photomicrograph of bioturbated, argillaceous, algal-crinoid-bryoaoan wackestone Crest Exploration Company Caldwell 2-1100 (depth, 3816 ft.).

(White bar in the lower right corner of photomicrographs represents 1 mm)

fusulinids, and endothyrid foraminifera. This rock type is present at the base of the Council Grove B-zone and is gradational with the dark-gray, bioturbated, argillaceous, crinoid-bryozoan wackestones below and the oolite grainstones above (Asquith, et al., 1977).

Oolite grainstone. — light-tan, cross-stratified oolite grainstones (Fig. 5-5) with oomoldic porosity (see: sonic log, Fig. 5-1). Other allochems include minor (<5 percent) fusulinids, brachiopod fragments, and gastropods (Asquith, et al., 1977).

Argillaceous, algal-crinoid-bryozoan wackestone. — medium to dark-gray, bioturbated wackestones with dasycladacean green algae, crinoids and bryozoans in an argillaceous, micrite matrix (Fig. 5-6).

Other fossil fragments include minor fusulinids, echinoid spines, brachiopod fragments and horn corals. This rock type overlies the Council Grove B-zone (Asquith, et al., 1977).

DEPOSITIONAL HISTORY

Modern oolite shoals on the Bahama platform (Purdy, 1961, p. 61; Hine, 1977, p. 1560) have a vertical sequence ranging from oolitic mud bearing facies (i.e. oolitic wackestones) to oolite facies (i.e. oolite grainstones). The Council Grove B-zone oolite shoal has a similar vertical sequence of bioturbated, oolitic-crinoid-bryozoan wackestones interbedded with thin oolite grainstones which grade upward into an oolite grainstone facies (Figs. 5-1 and 5-2). These vertical sequences represent progressively shallowing conditions as the result of oolite shoal progradation.

Overlying the oolite grainstone facies (Figs. 5-1 and 5-2) of the Council Grove B-zone are bioturbated, argillaceous, algal-crinoid-bryozoan wackestones which signify a return to quiet water deposition. The return to quiet water deposition occurred because of either a transgression of the sea or a shifting of the bi-directional tidal currents. These bi-directional tidal currents are necessary for oolite formation (Loreau and Purser, 1973, p. 321).

The isopach map of the Council Grove B-zone oolitic facies (Fig. 5-7) illustrates that the Council Grove oolite shoal has a depositional strike approximately east-west. The shoal has a long dimension of approximately 18 miles and a short dimension of approximately 11 miles with a maximum thickness of 60 feet (Fig. 5-7). Loreau and Purser (1973, p. 32) noted that oolites are commonly distributed over a wider area than their environment of origin. The areal extent, therefore, of the Council Grove oolite shoal does not necessarily represent the extent of the oolite forming environment.

Superimposed on the east-west trending depositional strike of the oolite shoal are a series of north-south trending isopach thicks and thins (Fig. 5-7). These thicks and thins make-up a ridge and swale system transversely oriented across the oolite shoal that formed because of north-south flowing bi-directional tidal currents. Similar ridge and swale systems have been noted across modern oolite shoals (Hoffmeister, et al., 1967; Ball, 1967). Examination of Figure 5-7 reveals a lobate outline on the south side of the Council Grove oolite shoal. The lobes probably represent spillover lobes like those present in modern oolite shoals (Hine, 1977, p. 1562).

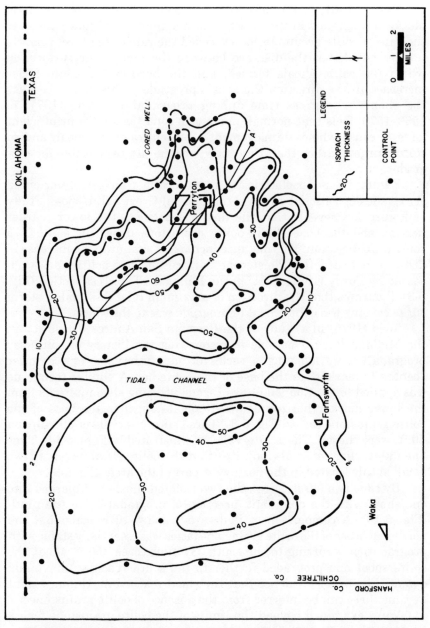

FIG. 5-7: Isopach map of the Lower Permian Council Grove B-zone, Ochiltree County, Texas (arrow indicates location of the Crest Exploration Company Caldwell 2-1100). (after: Asquith, et al., 1977).

Purdy (1961, p. 60) observed that oolite shoals of the Bahama platform prograded in the direction of maximum tidal flow whenever the rate of oolite accumulation exceeded the rate of sea level rise. He further noted that the distance between the Pleistocene surface, on which the oolite shoals formed, and the base of the oolite facies increased in the direction the shoals prograded. This occurs because the shoals transgress time during progradation. Hine (1977, p. 1569–1570) feels that normal, non-storm conditions have insufficient force to cause wholesale movement of an entire oolite shoal, and so storm surges, rather than tidal flow, are mainly responsible for progradation.

Figure 5-8 is an electric log cross-section transversely oriented across the depositional strike of the Council Grove oolite shoal. Study of Figure 5-8 reveals that the distance between the lower marker horizon and the base of the oolite grainstone facies increases in a southward direction. This distance increase is similar to the increase in distance between the Pleistocene surface and the oolitic facies observed by Purdy (1961, p. 61) in the Bahamas. The cross-section (Fig. 5-8) illustrates that the direction of maximum tidal flow and/or storm surge causing the oolite shoal to prograde was to the south.

Todd (1976) in a study of the Permian San Andres Formation in the Midland Basin was able to reconstruct the history of oolite-bar progradation by mapping a series of slice intervals from a lower "bentonite marker" to the base of the oolite-bar. A similar technique was applied to the Council Grove B-zone utilizing slice intervals from the lower marker horizon below the oolite shoal to the base of the oolite grainstone facies (Fig. 5-8). Three slice intervals of 30, 40, and 50 ft. were chosen. These intervals are illustrated in Figure 5-8. Time one (30-ft. slice interval) indicates that the oolite shoal began as two small shoals located in the northwest part of the area (Fig. 5-9).

By time two (40-ft. slice), the two isolated shoals had merged into one shoal with the dominant direction of progradation to the south (Fig. 5-10). A comparison of figures 5-9 and 5-10 reveals that the northern border of the oolite grainstone facies was essentially stable with progradation occurring to the south. By time three (50-ft. slice), the oolite shoal had prograded further to the south reaching maximum extent (Fig. 5-11). Evidence that the shoal reached maximum extent by time three can be inferred from the absence of oolite grainstones.

Wells examined beyond the limits of the oolite grainstone facies, illustrated in Figure 5-11, contained either bioturbated, oolitic-crinoid-bryozoan wackestones or bioturbated, argillaceous, crinoid-bryozoan wackestones, but the wells contained no oolite grainstones.

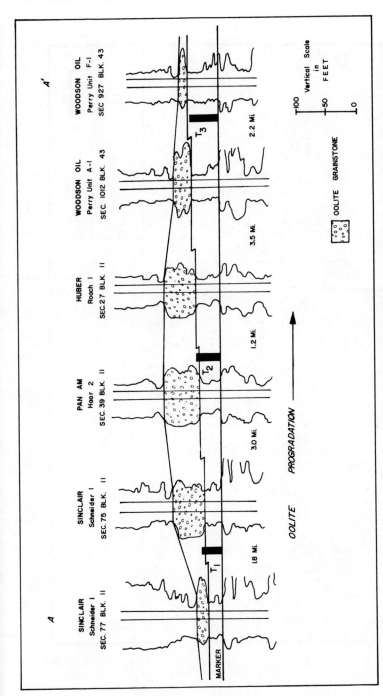

FIG. 5-8: Electric log cross-section of the Lower Permian Council Grove B-zone (spontaneous potential curve on the left and formation resistivity curve on the right). The vertical black bars on the marker horizon indicate slice intervals mapped above the marker (see: Figs. 5-9, 5-10, 5-11). The line of cross-section A-A' is indicated on Figure 5-7). (after: Asquith, et al., 1977).

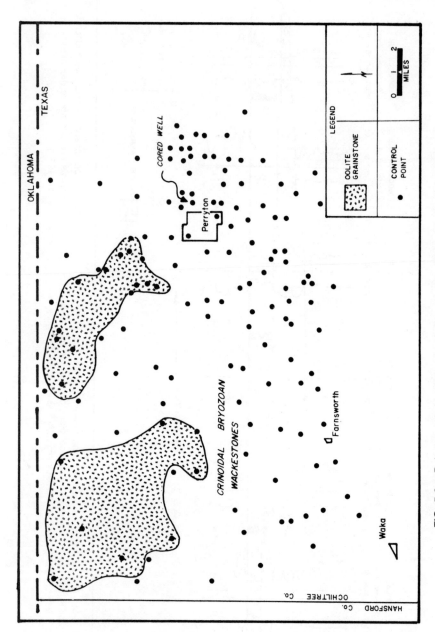

FIG. 5-9: Paleogeographic map of oolite grainstone facies for the Council Grove B-zone at time T₁ (30 ft. above the marker). (after: Asquith, et al., 1977).

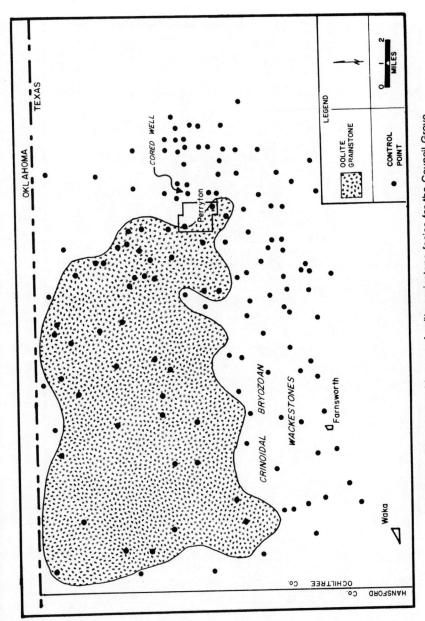

FIG. 5-10: Paleogeographic map of oolite grainstone facies for the Council Grove B-zone at time T_2 (40 ft. above the marker). (after: Asquith, et al., 1977).

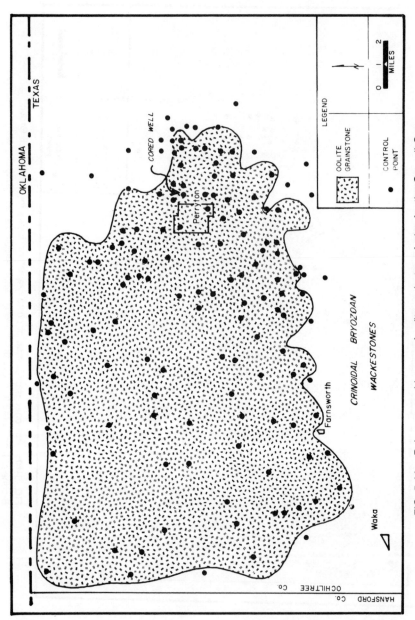

FIG. 5-11: Paleogeographic map of oolite grainstone facies for the Council Grove B-zone at time T₃ (50 ft. above the marker). (after: Asquith, et al., 1977).

The presence of bioturbated, argillaceous, algal-crinoid-bryozoan wackestones (Figs. 5-1 and 5-2) overlying the oolite grainstones signifies the return of quiet water deposition. Oomoldic porosity, in most of the wells examined petrographically, was noted in the oolite grainstone facies (Fig. 5-5). This indicates that, after deposition of the oolite shoal, parts of the shoal were subjected to fresh water leaching and diagenesis before the deposition of the overlying bioturbated, argillaceous, algal-crinoid-bryozoan wackestone (Asquith, et al., 1977).

REFERENCES

→Asquith, G. B., Russell, S. R., and Drake, J. E., 1977, Depositional history of the Lower Permian Council Grove B-zone oolite shoal, Ochiltree County, Texas: *The Compass*, v. 54, no. 2, p. 20–28.

Ball, M. M., 1967, Carbonate sand bodies of Florida and the Bahamas: *Jour. Sedimentary Petrology*, v. 37, no. 2, p. 556–591.

Blatt, Harvey, Middleton, G. E., and Murray, R. C., 1972, *Origin of Sedimentary Rocks:* Englewood Cliffs, N. J., Prentice-Hall, 634 p.

Hine, A. C., 1977, Lily Bank, Bahamas, history of an active oolite sand shoal: *Jour. Sedimentary Petrology*, v. 47, no. 4, p. 1554–1582.

Hoffmeister, J. E., Stockmans, K. W., and Multer, H.G., 1967, Miami Limestone of Florida and its recent Bahamian counterpart: *Geol. Soc. America, Bull.*, v. 78, p. 175–190.

Howell, J. V., ed., 1962, Glossary of Geology and Related Sciences: 2nd ed.: Washington, D.C., *The Am. Geological Inst.*, 325 p.

Leverson, A. I., 1961, *Geology of Petroleum:* San Francisco, W. H. Freeman and Company, 724 p.

Loreau, J. P., and Purser, B. H., 1973, Distribution and ultrastructure of Holocene ooids in the Persian Gulf, p. 279–328 *in* Purser, B. H., *ed., The Persian Gulf, Holocene carbonate sedimentation and diagenesis in a shallow epicontinental sea:* Berlin, Heidelberg, and New York, Springer-Verlag, 471 p.

Purdy, E. G., 1961, Bahamian oolite shoals, p. 53–62 *in* Peterson, J. A. and Osmond, J. C., *eds., Geometry of Sandstone Bodies:* Am. Assoc. Petroleum Geologists, 240 p.

→Todd, R. G., 1976, Oolite-bar progradation, San Andres Formation, Midland Basin: *Am. Assoc. Petroleum Geologists, Bull.*, v. 60, p. 907–925.

SUGGESTED READING

Oolite Shoals
Hine, A. C., 1977, Lily Bank, Bahamas: History of an active oolite sand shoal: *Jour. Sedimentary Petrology*, v. 47, no. 4, p. 1554–1582.

Loreau, J. P., and Purser, B. H., 1973, Distribution and ultrastructure of

Holocene ooids in the Persian Gulf, p. 279–328 *in* Purser, B. H., ed., *The Persian Gulf, Holocene carbonate sedimentation and diagenesis in a shallow epicontinental sea:* Berlin, Heidelberg, and New York, Springer-Verlag, 471 p.

Todd, R. G., 1976, Oolite-bar progradation, San Andres Formation, Midland Basin: *Am. Assoc. Petroleum Geologists, Bull.,* v. 60, p. 907–925.

6

Reef and Bank Models

General

Reefs and banks are often important petroleum reservoirs because of their topographic relief and high indigenous organic content (Link, 1950, p. 286). In North America, reefs and banks are the reservoirs in three fields classified as "giants": Swan Hills reefs in Alberta, Canada (Hemphill, et al., 1970), Horseshoe "atoll" in West Texas (Vest, 1970), and Rainbow-Zama reefs in northern Alberta, Canada (Barss, et al., 1970).

Wilson (1975, p. 20) capably outlines the problems of terminology that arise when discussing reefs and banks. Since terminology controversy exists, the reader should recognize that the following terms are defined according to this writer's application:

Reef—a ridge or moundlike structure built of inplace organisms that have the potential to act as wave-resistant framebuilders.

Bank — a non-wave-resistant ridge or moundlike structure built of non-framebuilding organisms.

Bioherm—a massive, unlayered, mound-shaped body in discordant relationship to the surrounding layered facies which drap over it. Thus, a bioherm can be either a reef or a bank. The term only implies shape.

Biostrome — a tabular-shaped, massively bedded carbonate grading gradually into the surrounding better layered facies. A biostrome, like a bioherm, can be either a reef or a bank; the term only implies shape.

Atoll—a ringlike reef encircling a central lagoon.

Fringing reef—a linear reef that trends parallel to the coastline but without a lagoon separating it from land.

Barrier reef—a linear reef that trends parallel to the coastline, and is separated from it by a wide lagoon or carbonate shelf.

Pinnacle reef—a small pyramid or cone-shaped reef usually less than four miles in diameter and several hundred feet in height. Besides pinnacle reefs, there are also pinnacle-shaped banks. Pinnacle, therefore, is a term denoting shape.

Mound—an equidimensional or ellipsoidal buildup (Wilson, 1975, p. 21).

It is apparent from the above definitions of a reef and a bank, that the knowledge of whether or not a particular organism had the potential for building a wave-resistant framework is important in environmental interpretation. Table 1 is a list of some of the major organisms categorized into framebuilders and non-framebuilders. The non-framebuilders are further subdivided into sediment binders and sediment bafflers (Table 1).

Modern reefs grow in shallow tropical seas usually in less than 100 ft. of water where salinity varies from 27-40 parts per thousand and the temperature rarely is lower than 20°C (Shepard, 1963, p. 351). An exception to these conditions is the coral reefs off the Norwegian coast that grow in colder water up to 200 feet deep (Teichert, 1958). The fauna of modern reefs is very complex. The reef framework is composed of corals, hydrocorallines, bryozoans, and calcareous red algae. Other organisms found associated with reefs include calcareous sponges, foraminifera, echinoids, pelecypods, and gastropods.

A reef complex can be broadly subdivided into three facies: (1) forereef, (2) reef flat (organic framework or reef core), and (3) back-reef (lagoon or shelf). The forereef is adjacent to the reef flat and consists of reef talus sloping seaward (Fig. 6-1). The reef talus is composed of reef debris eroded from the front of the reef flat. The forereef deposits become thinner and finer grained seaward where they grade into fine-grained, argillaceous basin deposits. The reef flat is made up of the wave-resistant framework of the reef complex and is often referred to as the reef core.

It is interesting to note that even though the reef framework or core is the most important part of the reef ecologically, it seldom makes up more than 10 percent of the total reef complex (Jardine, et

TABLE 1
CLASSIFICATION OF ORGANISMS BASED ON THEIR CONTRIBUTION TO BIOHERM DEVELOPMENT

FRAMEBUILDERS	NON-FRAMEBUILDERS	
	Sediment Binders	Sediment Bafflers
Massive and Tabular	Blue-green Algae	Fenestrate Bryozoan
Stromatoporoids	Encrusting Red Algae	Phylloid Algae
Massive Tabulate Corals	Tubular Forams	Dendroid Stromatoporoids
Hexacorals	Tubiphytes	Dendroid Corals
Hydrozoans	Encrusting Bryozoan	Branching Red Algae
Rudists		Branching Bryozoan
Calcareous Sponges		Segmented Green Algae
Richtofenid Brachiopods		Tetracorals
		Crinoids

al., 1977, p. 878). The ecologic importance of the reef core stems from its control on other facies (i.e. forereef and lagoon). The lagoon deposits or backreef shelf deposits (depending on scale) consist of fine-grained reef debris, lime mud, and often dolomite and evaporites. In the section on ancient reef models, the reader should note how each of these facies is interpreted from the carbonate rock types.

Banks can be broadly subdivided into two types: (1) mechanical accumulations and (2) sediment baffling and binding accumulations. Mechanical accumulations consist of banks built by wave and current transported debris. Sediment baffling and binding banks are built by the trapping of sediment by sediment baffling and binding organisms. Some banks are built by a combination of these two processes (Wilson, 1975, p. 165). Wilson (1975, p. 366–368) reported that when banks composed of sediment binding and baffling organisms (Table 1) reach wave-base, framebuilding organisms often colonize the top of the bank forming a crestal, wave-resistant, reef boundstone.

Porosity development in both reefs and banks is very complex because it can be of primary or secondary origin. The primary porosity includes both intergranular or organic porosity. Because of their topographic expression, reefs and banks are commonly areas of secondary porosity development due to vadose leaching or dolomitization. Besides primary porosity, leaching, and dolomitization, fracturing is also common.

Figure 6-2 illustrates the relative size and shape of eight well-known fossil reef and bank deposits. Also, included in Figure 6-2 are the barrels of oil (BOPA) or thousands of cubic feet of gas (MCFGPA) per acre for these carbonate reefs and banks.

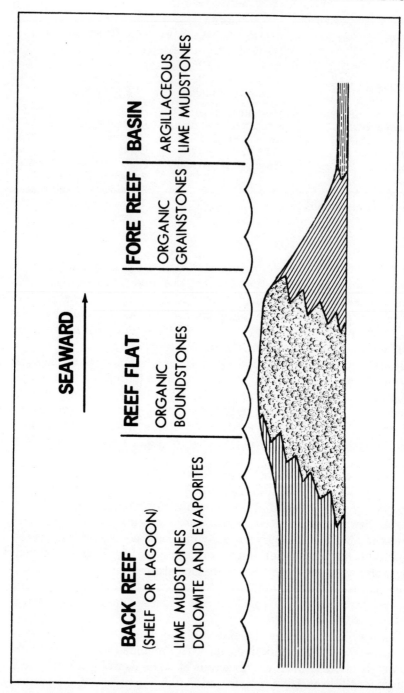

FIG. 6-1: Generalized cross-section illustrating the different facies across a reef complex.

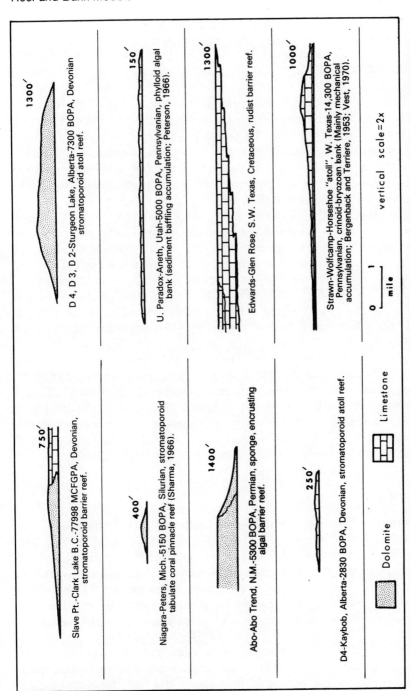

FIG. 6-2: Typical Reef and Bank Profiles (modified after: Achauer, C. W., 1968, ARCO Carbonate Seminar Notes; courtesy of the Atlantic Richfield Company).

1300′

D 4, D 3, D 2-Sturgeon Lake, Alberta-7300 BOPA, Devonian stromatoporoid atoll reef.

150′

U. Paradox-Aneth, Utah-5000 BOPA, Pennsylvanian, phylloid algal bank (sediment baffling accumulation; Peterson, 1966).

1300′

Edwards-Glen Rose, S.W. Texas, Cretaceous, rudist barrier reef.

1000′

Strawn-Wolfcamp-Horseshoe "atoll", W. Texas-14,300 BOPA, Pennsylvanian, crinoid-bryozoan bank (Mainly mechanical accumulation; Bergenback and Terriere, 1953; Vest, 1970).

750′

Slave Pt.-Clark Lake B.C.-77998 MCFGPA, Devonian, stromatoporoid barrier reef.

400′

Niagara-Peters, Mich.-5150 BOPA, Silurian, stromatoporoid tabulate coral pinnacle reef (Sharma, 1966).

1400′

Abo-Abo Trend, N.M.-5300 BOPA, Permian, sponge, encrusting algal barrier reef.

250′

D4-Kaybob, Alberta-2830 BOPA, Devonian, stromatoporoid atoll reef.

vertical scale=2x

0   1
|————|
mile

Dolomite

Limestone

Devonian Reefs of
Alberta, Canada

INTRODUCTION

The Upper Devonian Swan Hills (D-4) and Leduc (D-3) reefs are members and formations within the Beaverhill Lake Formation and Woodbend Group (Fig. 6-3). These reefs developed on carbonate platforms (biostromes) located west, northwest, and southeast of Edmonton, Alberta, Canada (Figs. 6-4 and 6-5). Off the platforms, thinner basinal deposits of dark, calcareous shale and argillaceous limestone accumulated (Figs. 6-4 and 6-5).

The Swan Hills (D-4) and Leduc (D-3) reefs are circular-shaped bioherms that can be classified as either atolls (Langton and Chin, 1968) where they are broader and have extensive lagoonal facies (Redwater and Sturgeon Lake; Fig. 6-5) or as pinnacle reefs (Langton and Chin, 1968) where they are narrower and have only minor lagoonal facies (Golden Spike; Fig. 6-5). While most of the reefs are circular, exceptions are: the fringing reefs around the Peace River Arch (Figs. 6-4 and 6-5) and the Rimby-Meadowbrook barrier reef complex (Fig. 6-5).

The platform (biostromes) attained a thickness of 100 ft. during Beaverhill Lake time (Hemphill, et al., 1970, p. 73) and up to 200 ft. during Woodbend time. The bioherms (reefs) built on these platforms vary from 160 to 1000 ft. in relief (Selley, 1970, p. 168 and 169), and are lens-shaped bodies with marginal slopes of ¾ degree to 5 degrees (Wilson, 1975, p. 129). Unlike the limestone Swan Hills reefs, most of the Leduc reefs are extensively dolomitized (Andrichuk, 1958).

PETROGRAPHY

General The biostromal and biohermal carbonates of the Beaverhill Lake Formation and the Woodbend Group (Fig. 6-3) and their associated basinal facies can be subdivided into the following rock types: (1) stromatoporoid boundstone, (2) stromatoporoid-intraclastic packstone, (3) amphipora wackestone, (4) peloidal packstone, (5) laminated, peloidal packstone, and (6) argillaceous, crinoid-brachiopod wackestone and calcareous, crinoid-brachiopod shale.

The stromatoporoid boundstones, amphipora wackestones, and the stromatoporoid-intraclastic packstones are the carbonate rock types comprising the platform (biostrome) and reef (bioherm) facies;

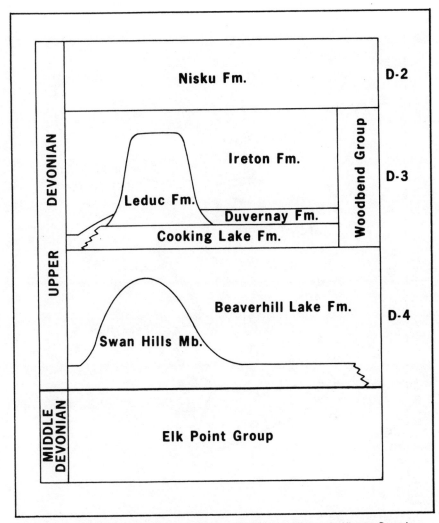

FIG. 6-3: Generalized stratigraphic column for the D-4 and D-3 reefs Alberta, Canada. (In part after: Thomas, 1962). From the Bulletin of the American Association of Petroleum Geologists, courtesy of the American Association of Petroleum Geologists.

the peloidal packstones and laminated, peloidal packstones represent the lagoonal facies, and the argillaceous, crinoid-brachiopod wackestones and calcareous, crinoid-brachiopod shales make-up the basinal facies. Detailed descriptions of the facies of Devonian reefs can be found in Klovan (1964), Thomas (1962), Jenik and Lerbekmo (1968), and Hemphill, et al. (1970).

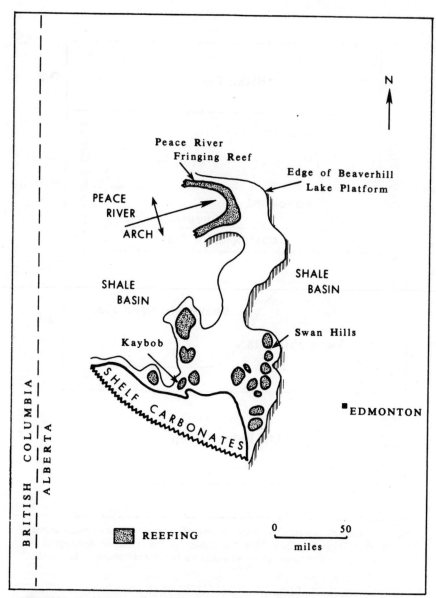

FIG. 6-4: Generalized paleogeographic map of Beaverhill Lake (D-4) deposition Alberta, Canda (Compiled from: Thomas, 1962, Jenik and Lerbekmo, 1968, and Hemphill, et al., 1970). The fringing reef associated with the Peace River arch is also referred to as the Springburn Reef (Hemphill, et al., 1970, p. 58).

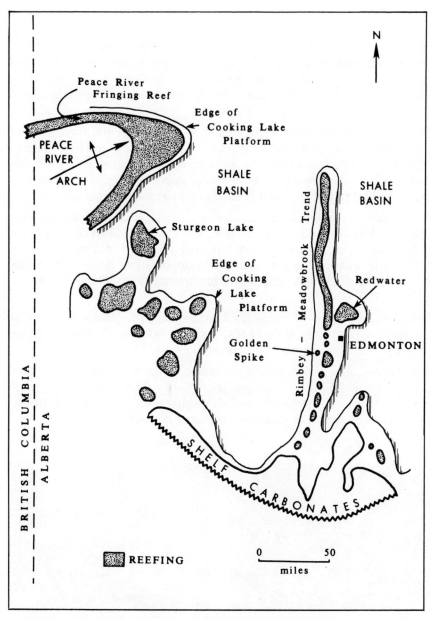

FIG. 6-5: Generalized paleogeographic map of Woodbend (D-3) deposition Alberta, Canada (compiled from: Andrichuk, 1958; Belyea, 1964, and Klovan, 1964).

Detailed Descriptions *Stromatoporoid boundstone.* — massive and tabular stromatoporoids and tabulate corals in a dark-brown, micrite matrix (Fig. 6-6). The micrite matrix contains minor amounts of amphipora along with fragments of brachiopods, crinoids, and rugose corals. The stromatoporoid boundstones are present in both the biohermal and the biostromal facies and represent the wave-resistant reef core facies.

Stromatoporoid-intraclastic packstone. — amphipora and fragments of stromatoporoids, and intraclasts in a dark-brown, micritic matrix (Fig. 6-7). The stromatoporoid-intraclastic packstones are reef talus and are best developed in the forereef deposits (Fig. 6-12), but can also be found in the backreef immediately behind the reef core.

Dolomitized, amphipora wackestone. — amphipora in a brown, micrite matrix (Fig. 6-8). In Figure 6-8, the micrite matrix has been dolomitized, and the amphipora have been leached and subsequently filled with calcite. Amphipora wackestones were deposited in the backreef lagoonal facies and in the biostromal facies (Hemphill, et al., 1970, p. 75).

Peloid packstones. — brown, peloid packstones with minor amounts of amphipora, ostracods, pelecypods, and calcispheres (Fig. 6-9). These peloid packstones are backreef lagoonal deposits.

Dolomitized, laminated, peloid packstone. — light-brown, dolomitized, laminated, peloidal packstones (Fig. 6-10). The fauna consists of minor ostracods, gastropods, and calcispheres. The laminated, peloid packstones comprise the backreef lagoonal deposits.

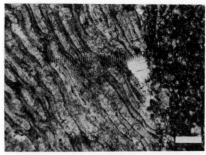 

FIG. 6-6: Photomicrograph of a stromatoporoid boundstone from the reef core. This same rock type is also found in the platform (biostromal) facies.

FIG. 6-7: Photomicrograph of stromatoporoid-intraclastic packstone which represents forereef talus.

(White bar in the lower right corner of photomicrographs represents 1 mm)

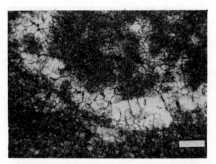
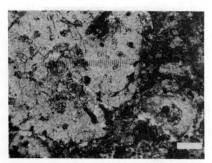

FIG. 6-8: Photomicrograph of amphipora wackestone which has been dolomitized. This sample is from the backreef lagoon (subtidal). Similar rock types can be found in the platform (biostromal) facies.

FIG. 6-9: Photomicrograph of peloidal packstone from backreef lagoonal facies.

(White bar in the lower right corner of photomicrographs represents 1 mm)

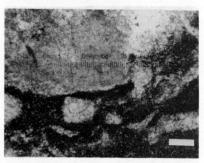

FIG. 6-10: Photomicrograph of dolomitized, laminated peloidal packstone from the backreef lagoonal facies.

FIG. 6-11: Photomicrograph of argillaceous, crinoid-brachiopod wackestone which represents deep water basinal facies adjacent to the biostromal and biohermal facies.

(White bar in the lower right corner of photomicrographs represents 1 mm)

Argillaceous, crinoid-brachiopod wackestones and calcareous, crinoid-brachiopod shales.— crinoid and brachiopod fragments in dark to light-gray, argillaceous, micritic or calcareous shale matrix (Fig. 6-11). These rock types were deposited in the deeper water basin areas adjacent to the carbonate bioherms and biostromes and are also post-reefing basinal fill deposits.

DEPOSITIONAL HISTORY

Figure 6-12 is a cross-section illustrating the major sedimentary facies and faunal distribution of a Devonian Swan Hills or a Leduc reef. As discussed earlier, the classification of these reefs as either atolls or pinnacle reefs is dependent both on their lateral extent and the development of the associated lagoonal facies (Langton and Chin, 1968). Figures 6-4 and 6-5 illustrate that both atolls and pinnacle reefs grew during Beaverhill Lake and Woodbend times.

Figure 6-13 is an isopach map of the Swan Hills Member (bioherm and biostrome facies) of the Beaver Hill Lake Formation (Fig. 6-3). The Swan Hills stromatoporoid-amphipora biostrome (platform) developed on the evaporites of the Middle Devonian Elk Point Group (Thomas, 1962, p. 211). Superimposed on these carbonate platforms are numerous (Fig. 6-13) circular-shaped, stromatoporoid-amphipora bioherms (reefs) that grew to heights of up to 350 ft. (Hemphill, et al., 1970, p. 73). In deeper water around the carbonate platforms, argillaceous, crinoid-brachiopod wackestones and calcareous, crinoid-brachiopod shales accumulated.

The basinal sedimentation, however, did not keep pace with reef growth, and, consequently, most of the argillaceous, crinoid-brachiopod wackestones and calcareous, crinoid-brachiopod shales of the Beaverhill Lake Formation represent post-reefing basin fill. These stratigraphic relationships are illustrated in Figure 6-14 by an electric log cross-section across the Swan Hills reef. The dark line in Figure 6-14 outlining the bioherm and biostrome facies of the Swan Hills Member is a time line. For a detailed account of the depositional history of the Swan Hills reefs see: Hemphill, et al. (1970).

Following deposition of the Beaverhill Lake Formation, there was a transgression of the sea, and the area of shelf carbonates was displaced to the south and southeast (Figs. 6-4 and 6-5). Like Beaverhill Lake deposition, the carbonates of the Woodbend Group began accumulating as stromatoporoid-amphipora platforms (biostromes) called the Cooking Lake (Thomas, 1962, p. 210).

These platforms grew to a thickness of 200 feet, and on these platforms, the circular-shaped, stromatoporoid-amphipora reefs (Leduc Formation) developed reaching heights of up to 1000 ft. (Selley, 1970, p. 169). In deeper parts of the basin away from the reefs dark, bituminous shale and argillaceous limestones of the Duvernay Formation (Thomas, 1962, p. 210) accumulated. Following reef growth, the deep areas between the reefs were filled with calcareous, crinoid-brachiopod shales of the Ireton Formation.

Jardine, et al. (1977) noted in a study of three Canadian Devonian reefs (Clark Lake, Redwater, and Judy Creek) that porosity distribution was predominantly controlled by depositional facies. They also found that, in addition to organic and intergranular poros-

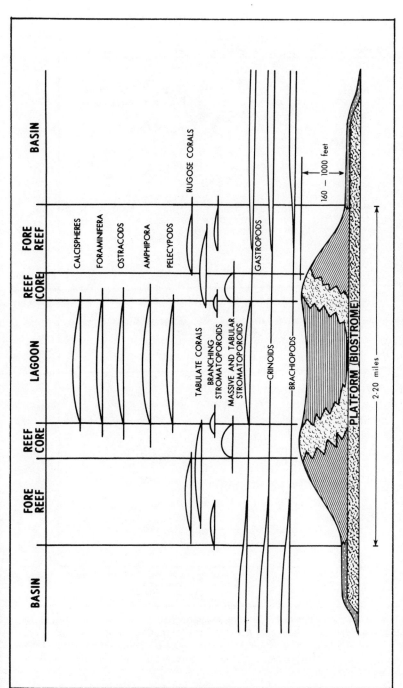

FIG. 6-12: Generalized cross-section across a Canadian Devonian reef (Paleontology compiled after: Andrichuk, 1958, Jenik and Lerbekmo, 1968, and Klovan, 1964).

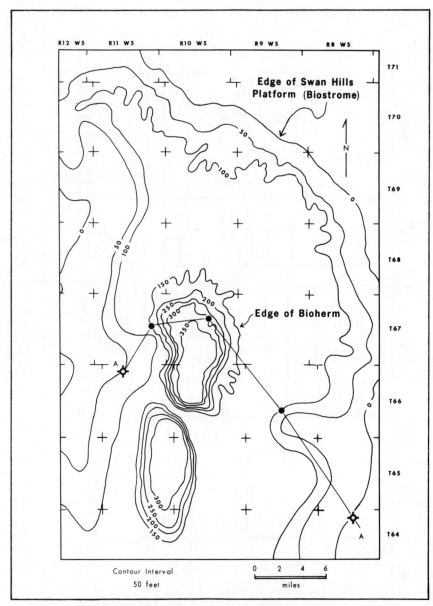

FIG. 6-13:   Isopach map of the Swan Hills Member (biohermal and biostromal facies) of the Beaverhill Lake Formation Alberta, Canada (modified after: Hemphill, et al., 1970). A-A′ line of cross-section illustrated in Figure 6-14. From the Bulletin of the American Association of Petroleum Geologist, courtesy of the American Association of Petroleum Geologists.

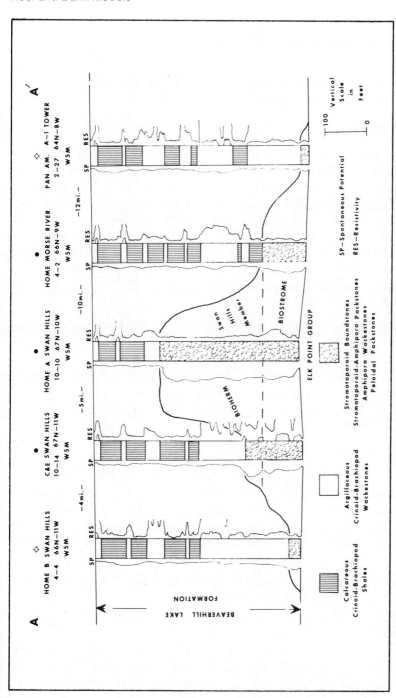

FIG. 6-14: Electric log cross-section across Swan Hill reef Alberta, Canada. Line of cross-section illustrated in Figure 6-13. The position of the different carbonate facies is based on the work of Hemphill, et al.. (1970). From the Bulletin of the American Association of Petroleum Geologists, courtesy of the American Association of Petroleum Geologists.

ity, there was secondary porosity developed due to leaching, fracturing, and often dolomitization.

Edwards-Glen Rose Barrier Reef, Texas Gulf Coast

INTRODUCTION

The upper Edwards-Glen Rose barrier reef complex of the Texas Gulf Coast is of Lower Cretaceous (Albian) age. The barrier reef developed on the edge of a shallow water shelf (Fig. 6-15) nearly 200 miles from the near-shore sandstones of the Hensel Formation that are adjacent to the Llano Uplift. South and southeast of the reef front was a deep water basin (Griffith, et al., 1969, p. 131; Fisher and Rodda, 1969, p. 56). Griffith, et al., (1969) subdivided the barrier reef complex into a forereef basinal facies (Atascosa Group), a reefoidal facies (Stuart City Formation), and a back-reef shelf-lagoon facies (Edwards Formation).

The Edwards-Glen Rose reef complex is underlain by the Pearsall Formation and overlain by the Georgetown Formation. The Edwards barrier reef complex extends into Louisiana in the subsurface and outcrops in Mexico where it is called the El Abra Reef. This complex may have extended over a distance of 900 to 1000 miles (Griffith, et al., 1969, p. 121) during the Lower Cretaceous (Fig. 6-15).

PETROGRAPHY

General   The Edwards-Glen Rose barrier reef complex is subdivided, for the purposes of this discussion, into six rock types: (1) rudist-coral boundstone, (2) rudist grainstone, (3) rudist wackestone, (4) miliolid wackestone, (5) miliolid grainstone, and (6) globigerinal wackestone. Rudist-coral boundstones and rudist wackestones and grainstones make-up the reef and near-reef facies, and miliolid wackestones and grainstones make-up the backreef shelf-lagoon. Globigerinal wackestones were deposited in the forereef basin.

Detailed Descriptions   *Rudist-coral boundstone.* — in-situ caprinids and corals (Cladophyllum) with minor amounts of Eoradiolites, Toucasia, and Chondrodonta debris and micrite. Secondary porosity is the result of leaching of the caprinids (Fig. 6-16).

Rudist grainstone. — poorly sorted, rudist debris cemented with sparry calcite cement with minor micrite matrix (Fig. 6-17); additional fossil debris include Chondrondonta and coral fragments.

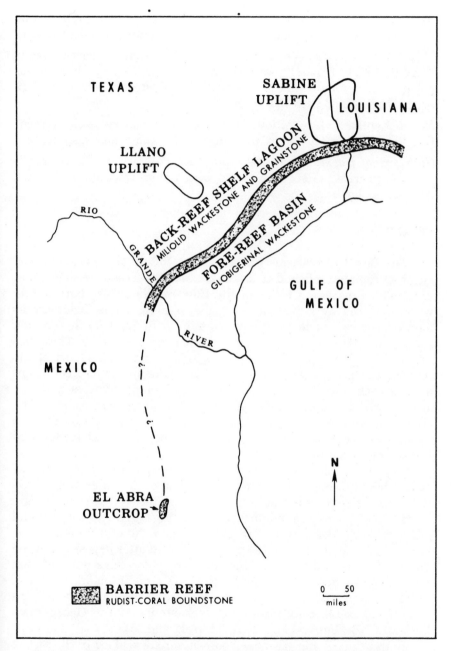

FIG. 6-15: Generalized paleogeographic map of south Texas, northern Mexico and Louisiana during deposition of the Lower Cretaceous Edwards-Glen Rose barrier reef. (The position of the barrier reef after: Griffith, et al., 1969).

Rudist wackestone. — rudist debris in a dark, micrite matrix (Fig. 6-18). Other fossils include Chondrondonta and minor Cladophyllum.

Miliolid wackestone. — miliolids in a dark, peloidal, micritic matrix (Fig. 6-19). Additional fossils include echinoid spines and minor pelecypod fragments. The numerous vugs illustrated in Figure 6-19 are partially filled with sparry calcite cement.

Miliolid grainstone. — miliolids cemented with sparry calcite cement (Fig. 6-20). Other allochems are minor oolites and echinoid spines.

Globigerinal wackestone. — planktonic foraminifera in a dark, argillaceous, micrite matrix (Fig. 6-21) with minor amounts of mollusk and echinoid debris.

DEPOSITIONAL HISTORY

The Lower Cretaceous Edwards-Glen Rose is a prograding barrier reef complex referred to as the Stuart City Trend by Griffith, et al. (1969). Prograding facies of the Edwards-Glen Rose barrier reef were described by Bebout and Loucks in 1974. These facies are as follows: The lower one-quarter or more of the Stuart City (lower Glen Rose) consists of dark-colored, planktonic foraminifera wackestones (Fig. 6-21); similar carbonates are present in the upper part of the section seaward of the barrier trend and make-up the forereef basinal facies. Open-marine, dark, planktonic foraminifera wackestones are followed by lower shelf-slope, echinoid-mollusk wackestones. The progradational sequence continues (Fig. 6-22) with gradation into upper shelf-slope, coral-stromatoporoid boundstones (patch reef) and rudist-coral wackestones (Fig. 6-18).

The upper shelf-slope carbonates are overlain (Fig. 6-22) by rudist boundstones (Fig. 6-16) and rudist grainstones (Fig. 6-17) which comprise the shelf-margin barrier reef facies. The shelf-margin barrier reef facies are in turn overlain (Fig. 6-22) by the miliolid grainstones and wackestones (Figs. 6-19 and 6-20) of the shelf-lagoon facies. A period of subaerial exposure occurred after deposition of the Stuart City as indicated by the presence of breccias and solution zones. Rapid foundering of the barrier reef trend brought an end to rudist growth and resulted in deposition of the dark-gray, open-marine, planktonic foraminifera wackestones of the Georgetown Formation (Bebout and Loucks, 1974, p. 12 and 15).

Both primary and secondary porosity are developed in the Stuart City (Edwards-Glen Rose) barrier reef. The primary porosity includes both intergranular and organic porosity, while secondary porosity

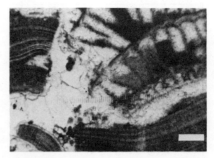

FIG. 6-16: Photograph of rudist-coral boundstone (reef core).
(Vertical white bar on left represents 1 cm)

FIG. 6-17: Photomicrograph of poorly sorted rudist grainstone cemented with sparry calcite cement (backreef).
(White bar in the lower right corner of photomicrographs represents 1 mm)

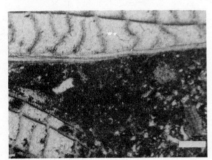
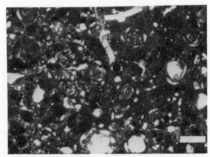

FIG. 6-18: Photomicrograph of rudist wackestone (found in both forereef and backreef deposits).

FIG. 6-19: Photomicrograph of miliolid wackestone (backreef shelf-lagoon).
(White bar in the lower right corner of photomicrographs represents 1 mm)

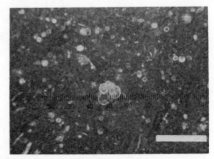

FIG. 6-20: Photomicrograph of miliolid grainstone (backreef shelf-lagoon).

FIG. 6-21: Photomicrograph of globigerinal wackestone (forereef basinal facies).
(White bar in the lower right corner of photomicrographs represents 1 mm)

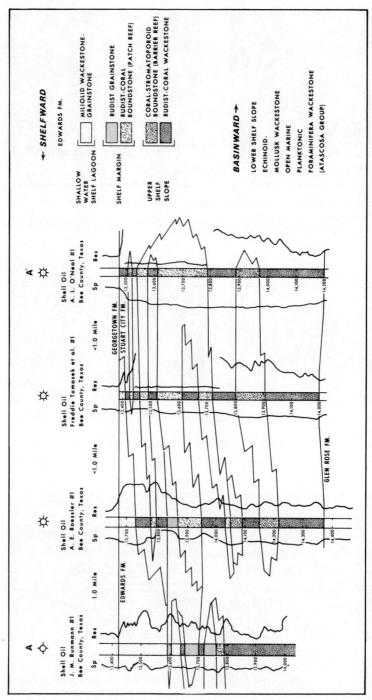

FIG. 6-22: Electric log cross-section across the Edwards-Glen Rose (Stuart City) barrier reef Bee County, Texas. The various carbonate facies modified after: Bebout and Loucks, 1974.

includes solution enlarged vugs, fractures, and moldic porosity. Most of the porosity occurs in amounts of less than five percent and in thin intervals (Bebout and Loucks, 1974, p. 72). Both primary and secondary porosity occur only in the rudist boundstone and grainstone facies; therefore, the vertical location of porosity encountered in a well is dependent on the position of the well relative to the shelf edge.

Because the barrier reef is regressive, wells located near the shelf edge have porosity developed near the top of the Stuart City, and wells located back toward the shelf lagoon have porosity developed progressively lower in the section (Bebout and Loucks, 1974, p. 72).

Mississippian Chappel Crinoidal-Bryozoan Bank, Conley Field, Hardeman County, Texas

INTRODUCTION

The Conley Field is located in Block H of the Waco and Northwestern Railroad Survey in Hardeman County, Texas (Fig. 6-23), in the central part of the Hardeman Basin. Along with Mississippian carbonate bank production, there is also production from the Pennsylvanian Palo Pinto Limestone and the Ordovician Arbuckle Formation.

The Mississippian Chappel crinoid-bryozoan bank is Osage and Meramec in age (Allison, 1979). The Chappel Bank lies unconformably on the Ordovician Arbuckle Formation, and it is overlain by the Mississippian St. Louis Limestone (Fig. 6-24). The best porosities are developed in areas where the Chappel Formation is thickest because of bank buildup (Fig. 6-24). The porous zones result from a combination of intercrystalline porosity due to dolomitization of crinoid-bryozoan wackestone fracturing and leaching. Dolomitization, however, is often irregular in distribution, and in the Conley Field, porosity in some of the wells results primarily from fracturing or leaching. The sonic prosity in these wells is low (<10 percent) and petrographic analysis reveals low matrix porosity. Wells with sucrosic (intercrystalline) porosity due to dolomitization commonly have sonic porosities greater than 10 percent.

PETROGRAPHY

General The carbonates of the Chappel Formation can be subdivided into four carbonate rock types: (1) crinoid-bryozoan grainstone, (2) crinoid-bryozoan wackestone, (3) argillaceous, crinoid-

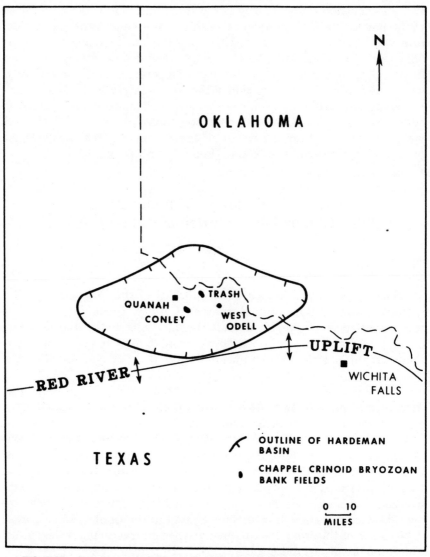

FIG. 6-23:   Location map of Conley Field Hardeman County, Texas.

bryozoan wackestone, and (4) dolomitized, crinoid-bryozoan wacke-stone.

Detailed Descriptions   *Crinoid-bryozoan grainstone.* — well-sorted crinoid, bryozoan, and minor brachiopod fragments with minor micrite

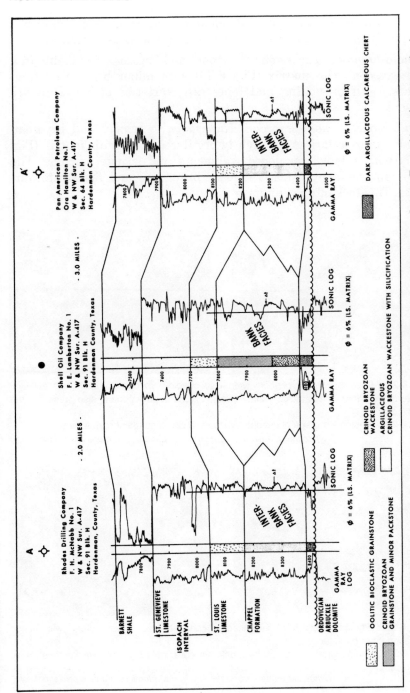

FIG. 6-24: Cross-section across the Chappel crinoid-bryozoan bank Conley Field Hardeman County, Texas. Line of cross-section is illustrated in Figure 6-31.

matrix cemented with epitaxial calcite cement (Fig. 6-25). The bryo-zoan fragments include both branching and fenestrate varieties.

Crinoid-bryozoan wackestone. — crinoid and bryozoan fragments in a light-gray, micrite matrix (Fig. 6-26) with minor brachiopod frag-ments. Both branching and fenestrate varieties of bryozoan are present.

Argillaceous, crinoid-bryozoan wackestone. — crinoid and bryozoan fragments in a dark-gray, slightly argillaceous, micritic matrix (Fig. 6-27) with minor brachiopod fragments and sponge spicules. The bryozoan fragments consist of both branching and fenestrate var-ieties. This rock often exhibits silification.

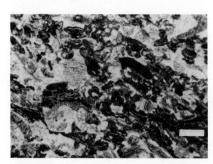 

FIG. 6-25: Photomicrograph of poorly winnowed, crinoid-bryozoan grainstone with epitaxial calcite cement (bank flank facies). Minor crinoid-bryozoan packstone is also present in the flank facies.

FIG. 6-26: Photomicrograph of crinoid-bryozoan wackestone (bank core facies).

(White bar in the lower right corner of photomicrographs represents 1 mm)

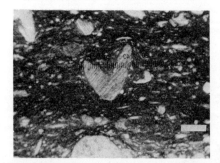 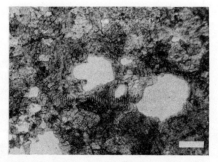

FIG. 6-27: Photomicrograph of argillaceous, crinoid-bryozoan wackestone with minor silification (inter-bank facies).

FIG. 6-28: Photomicrograph of leached, dolomitized, crinoid-bryozoan wackestone (bank core facies).

(White bar in the lower right corner of photomicrographs represents 1 mm)

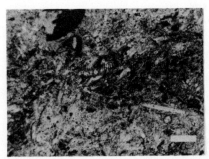 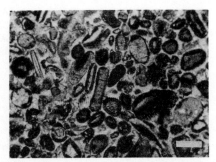

FIG 6-29: Photomicrograph of argillaceous, calcareous chert (offshore, open-marine facies).
FIG. 6-30: Photomicrograph of oolitic-bioclastic grainstone. This sample is from the Mississippian St. Louis Limestone, which is the capping grainstone facies overlying the Chappel banks.

(White bar in the lower right corner of photomicrographs represents 1 mm)

Dolomitized, crinoid-bryozoan wackestone. — crinoid and bryozoan fragments with the micrite matrix selectively replaced by xenotopic dolomite (dolomite crystals 300-600μ). The undolomitized fossil debris is frequently leached forming fossilmoldic porosity (Fig. 6-28).

Argillaceous, calcareous chert. — gray to dark-gray, argillaceous chert with minor, very fine-grained calcite. Rod-shaped sponge spicules (Fig. 6-29) and fine-grained pyrite are also present. Calcite is often found filling fractures. This rock type makes-up the basal Mississippian (Osage?) in the Hardeman Basin, and represents an off-shore deep water environment.

DEPOSITIONAL HISTORY

The crinoid-bryozoan carbonates of the Mississippian Chappel Formation accumulated on the eroded surface of the Ordovician Arbuckle Formation (Fig. 6-24). The Mississippian depositional sequence began with the deposition of the Osage? off-shore, open-marine, argillaceous, calcareous cherts which graded upward into the Chappel crinoid-bryozoan bank and adjacent inter-bank facies (Fig. 6-24). Table 2 is a listing of the petrographic variation between the Chappel Formation bank and inter-bank facies (Fig. 6-24).

The superposition of thicker sections of Chappel Formation (Fig. 6-24) on structural highs suggests that the bank buildups were initiated and controlled by paleotopographic (structural) highs. The bank

TABLE 2
PETROGRAPHIC VARIATION BETWEEN CRINOIDAL-BRYOZOAN BANK AND
INTER-BANK FACIES, MISSISSIPPIAN CHAPPEL FORMATION, HARDEMAN
BASIN, HARDEMAN COUNTY, TEXAS

Bank Facies	Inter-bank Facies
crinoid-bryozoan grainstone common	crinoid-bryozoan grainstone rare
crinoid-bryozoan wackestone commonly dolomitized and often exhibit fossilmoldic porosity	argillaceous crinoid-bryozoan wackestone common exhibits some silification
sponge spicules absent	sponge spicules present

buildups result from a combination of mechanical accumulation of crinoid-bryozoan grainstones and sediment baffling of micrite, forming crinoid-bryozoan wackestones (Wilson, 1975, p. 166–167). The fan-shaped fenestrate bryozoans probably acted as a dominant sediment baffling organism as they did in later Pennsylvanian bank deposits (Wermund, 1975, p. 27).

Wilson (1975, p. 166) has noted that crinoid-bryozoan wackestones make up the core of Mississippian banks and the associated crinoid-bryozoan grainstones represent the winnowed flank deposits. Wells, therefore, which penetrate a bank and encounter crinoid-bryozoan grainstones are in a flank position off the topographic crest of the bank. In the Conley Field, wells drilled off the crest of the Chappel Bank (Fig. 6-31) commonly penetrate thick sequences of crinoid-bryozoan grainstone (flank deposits). The presence of extensive flank deposits indicates that the top of the bank remained at wave-base for an extended period of time (Wilson, 1975, p. 369).

The common occurrence of crinoid-bryozoan grainstones and the presence of dolomitization and leaching in the bank facies (Table 2), indicates that the banks remained as topographic highs at or near wave-base during Chappel deposition. Many Chappel wells have more than one porosity zone (Fig. 6-24). Examination of cores and cuttings reveals that fossilmoldic porosity is often developed in the different porosity zones. The presence of the fossilmoldic porosity and solution breccia zones indicates that bank buildup was not continuous but was interrupted by periods of subaerial exposure and vadose leaching.

Wilson (1975, p. 367 and 369) suggests that when sea level remains stable, carbonate mound and intermound areas are often covered by a cap of cross-stratified grainstone. The capping

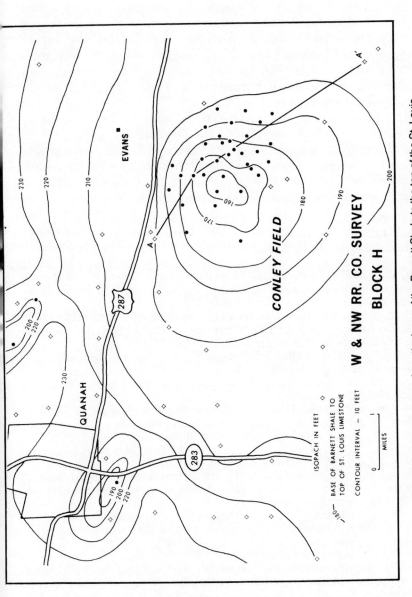

FIG. 6-31: Isopach map from the base of the Barnett Shale to the top of the St. Louis Limestone, illustrating thinning over the Chappel crinoid-bryozoan bank at the Conley Field Hardeman County, Texas. A-A' line of cross-section illustrated in Figure 6-24.

grainstone of the Chappel banks is the oolitic bioclastic St. Louis Limestone (Figs. 6-30 and 6-24).

The Chappel banks can be mapped in the subsurface by either isopaching the Chappel Formation, identifying isopach thicks, or isopaching the horizons immediately above the Chappel, looking for thinning over the bank buildups (Figs. 6-24 and 6-31). In the Hardeman Basin, the latter isopach method is used because some wells do not penetrate the Arbuckle Formation, and the Chappel-Arbuckle contact is often hard to pick.

Dip meters can also be used in exploration for Chappel banks by mapping the drap of units overlying the Formation. Once dip direction of the drap is determined, wells can be offset in the opposite direction to achieve a topographically higher position on the bank. Subsurface mapping techniques outlined for the Chappel banks are also applicable to the mapping of other bank and reef deposits.

REFERENCES

Allison, M. D., 1979, Petrology and depositional environment of the Mississippian Chappel bioherms Hardeman County, Texas: *M.S. thesis,* West Texas State University, Canyon, Texas, 45 p.

Andrichuk, J. M., 1958, Stratigraphy and facies analysis of Upper Devonian reefs in Leduc, Settler, and Redwater areas, Alberta: *Am. Assoc. Petroleum Geologists, Bull.,* v. 42, p. 1–93.

Barss, D. L., Copland, A. B., and Ritchie, W. D., 1970, Geology of Middle Devonian reefs, Rainbow area, Alberta, Canada, p. 18–49 *in* Halbouty, M. T., *ed., Geology of Giant Petroleum Fields:* Am. Assoc. Petroleum Geologists, Mem. 14, 575 p.

Bebout, D. G., and Loucks, R. G., 1974, Stuart City Trend Lower Cretaceous, South Texas, a carbonate shelf-margin model for hydrocarbon exploration: Univ. Texas, Austin, Bureau. Econ. Geology, *Rept. Inv. 78,* 80 p.

Belyea, H. R., 1964, Upper Devonian, pt. II *in* Geological History of Western Canada: *Alberta Soc. Petroleum Geologists,* p. 66–81.

Bergenback, R. E., and Terriere, R. T., 1953, Petrography and petrology of Scurry Reef Scurry County, Texas: *Am. Assoc. Petroleum Geologists, Bull.,* v. 37, no. 5, p. 1014–1029.

Fisher, W. L., and Rodda P. U., 1969, Edwards Formation (Lower Cretaceous), Texas, dolomitization in a carbonate platform system: *Am. Assoc. Petroleum Geologists,* v. 53, no. 1, p. 55–72.

Griffith, L. S., Pitcher, M. G., and Rice, G. W., 1969, Quantitative environmental analysis of a Lower Cretaceous reef complex, p. 120–138 *in* Friedman, G. M., *ed., Depositional environments in carbonate rocks, A symposium:* Soc. of Econ. Paleontologists and Mineralogists, Spec. Pub. No. 14, 209 p.

Hemphill, C. R., Smith, R. I., and Szabo, F., 1970, Geology of Beaverhill Lake reefs, Swan Hills area, Alberta, p. 50–90 *in* Halbouty, M. T., *ed., Geology of Giant Petroleum Fields:* Am. Assoc. Petroleum Geologists, Mem. 14, 575 p.

Jardine, D. E., Andrew, D. P., Wishart, J. W., and Young, J. W., 1977, Distribution and continuity of carbonate reservoirs: *Jour. Petroleum Technology,* no. 10, p. 873–885.

Jenik, A. K., and Lerbekmo, J. F., 1968, Facies and geometry of Swan Hills Reef Member of Beaverhill Lake Formation (Upper Devonian), Goose River Field, Alberta, Canada: *Am. Assoc. Petroleum Geologists, Bull.,* v. 52, p. 21–56.

Klovan, J. E., 1964, Facies analysis of the Redwater Reef complex, Alberta, Canada: *Canadian Petroleum Geology Bull.,* v. 12, p. 1–100.

Langton, J. R., and Chin, G. E., 1968, Rainbow Member facies and related reservoir properties, Rainbow Lake, Alberta: *Am. Assoc. Petroleum Geologists, Bull.,* v. 52, no. 10, p. 1925–1955.

Link, T. A., 1950, Theory of transgressive and regressive reef (bioherm) development and origin of oil: *Am. Assoc. Petroleum Geologists, Bull.,* v. 34, no. 2, p. 263–294.

Peterson, J. A., 1966, Stratigraphic vs. structural controls on carbonate-mound hydrocarbon accumulation, Aneth Area, Paradox Basin: *Am. Assoc. Petroleum Geologists, Bull.,* v. 50, no. 10, p. 2068–2081.

Selley, R. C., 1970, *Ancient sedimentary environments, a brief survey:* London, Chapman and Hall Ltd., 237 p.

Sharma, G. D., 1966, Geology of Peters Reef, St. Clair County, Michigan: *Am. Assoc. Petroleum Geologists, Bull.,* v. 50, no. 2, p. 327–350.

Shepard, F. P., 1963, *Submarine Geology,* 2nd ed.: New York, Evanston, London, Harper and Row, 559 p.

Teichert, C., 1958, cold- and deep-water coral banks: *Am. Assoc. Petroleum Geologists,* v. 42, p. 1064–1082.

Thomas, G. E., 1962, Grouping of carbonate rocks into textural and porosity units for mapping purposes, p. 193–223 *in* Ham, W. E., *ed., Classification of carbonate rocks, a symposium:* Am. Assoc. Petroleum Geologists, Mem. 1, 279 p.

Vest, E. L., Jr., 1970, Oil fields of Pennsylvanian-Permian Horseshoe atoll, West Texas, p. 185–203 *In* Halbouty, M. T., *ed., Geology of Giant Petroleum Fields:* Am. Assoc. Petroleum Geologists, Mem. 14, 575 p.

Wermund, E. G., 1975, Upper Pennsylvanian Limestone Banks, North Central Texas: Univ. Texas, Austin, Bur. Econ. Geology, *circ.* 75–3, 34 p.

Wilson, J. L., 1975, *Carbonate facies in geologic history:* Berlin, Heidelberg, and New York, Springer-Verlag, 471 p.

SUGGESTED READING

Reefs and Banks

Fisher, J. H., 1977, Reefs and evaporites — concepts and depositional models: Am. Assoc. Petroleum Geologists, *Studies in Geology No. 5.,* 204 p.

Heckel, P.H., 1974, Carbonate buildups in the geologic record, a review, p. 90–154 *in* Laporte, L. F., *ed., Reefs in Time and Space, selected examples from the recent and ancient:* Soc. of Econ. Paleontologists and Mineralogists, Spec. Pub. No. 18, 256 p.

—————————, and O'Brien, G.D., 1975, Silurian reefs of Great Lakes Region of North America: *Am. Assoc. Petroleum Geologists, Bull., Reprint series No. 14,* 243 p.

Newell N. D., 1974, *The evolution of reefs in Planet Earth:* San Francisco, W. Freeman and Company, p. 182–194.

Perkins, B. F., 1975, Carbonate Rocks III, organic reefs: *Am. Assoc. of Petroleum Geologists, Bull., Reprint series No. 15,* 190 p.

Wermund, E. G., 1975, Upper Pennsylvanian Limestone Banks, North Central Texas: Univ. Texas, Austin, Bur. Econ. Geology, *Circ.* 75–3, 34 p.

Wilson, J. L., 1975, *Carbonate facies in geologic history:* Heidelberg, Berlin, and New York, Springer-Verlag, 471 p.

7

Lithology Logging

General

Mechanical logs provide most of the subsurface data used by exploration geologists. For this reason, it seems appropriate to conclude our discussion of subsurface carbonate models by describing the methods used to determine carbonate lithologies from logs. Four methods will be presented: (1) combination gamma ray CNL*-FDC* log, (2) M-N* cross-plot, (3) MID* cross-plot, and (4) Carbonate Rock Type Identification (Pickett, 1977). These four methods will be discussed in the order of their ease of application.

Combination Gamma Ray and CNL*-FDC* Log

The gamma ray log measures natural radioactivity of a formation. Shale-free carbonates (limestone or dolomite) and anhydrites are low in radioactive material and, consequently, give low gamma ray readings (Fig. 7-1). As the amount of shale increases, the gamma ray response increases (Fig. 7-1) because of the more radioactive nature of shales. CNL* logs (compensated neutron log) are obtained from a porosity tool that responds primarily to the amount of hydrogen in a formation.

FDC* logs (formation density compensated) are provided by a porosity tool that measures the electron density of a formation. A

*A Mark of Schlumberger

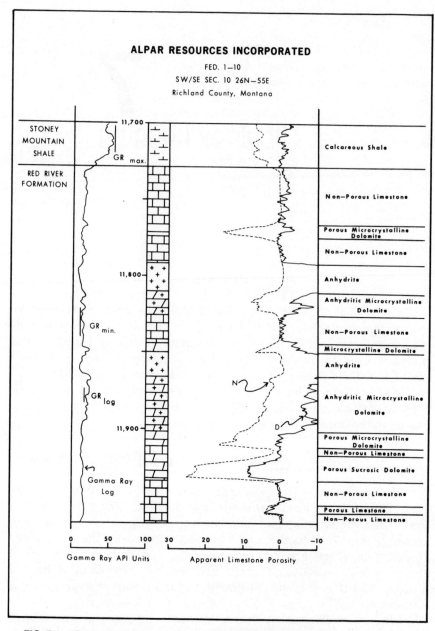

FIG. 7-1: Schlumberger combination gamma ray, compensated neutron, and formation density log Alpar Resources Federal 1-10 Sec. 10, T26N-R55E Richland County, Montana, illustrating the relationship of lithology to log responses. The lithologies for the interval were also checked by petrographic analysis of cuttings.

complete description of the gamma ray, CNL*, and FDC* logs is presented in the *Schlumberger Log Interpretation Manual, Volume I, Principles* (1972, p. 43–59). Figure 7-1 is a combination gamma ray and CNL*-FDC* log through the Ordovician Stony Mountain Shale and Red River Formation in Richland County, Montana and illustrates how lithologies are related to log response. As a "quick look" method when only a limited number of lithologies are present, this log package is satisfactory for basic lithologic mapping.

The gamma ray log may also be used to determine the amount (volume) of shale in carbonate rocks. Once the volume of shale is determined, carbonate/shale ratio maps can be constructed and used for carbonate facies analysis (Wermund, 1975, p. 9). The formula for determining the volume of shale is:

$$\text{Vsh} \quad (\text{in}\%) = \frac{\text{GR log} - \text{GR min}}{\text{GR max} - \text{GR min}}$$

Where: GR max = maximum gamma ray reading (shale)
GR min = minimum gamma ray reading (clean carbonate or sand)
GR log = gamma ray reading of formation

See Figure 7-1 for example
Schlumberger, 1974, Log Interpretation, vol. II/Applications, p. 50.

If a gamma-ray log is unavailable, the spontaneous potential (SP) log or neutron logs (CNL* or SNP*) can be used to calculate the volume of shale. The formulas for determining volume of shale from these logs are stated as follows:

$$\text{Vsh} \quad (\text{in}\%) = \frac{\text{SP} - \text{SPsh}}{\text{SPsd} - \text{SPsh}}$$

Where: SP = spontaneous potential of formation
SPsd = spontaneous potential (clean sand or carbonate)
SPsh = spontaneous potential (shale)

Schlumberger, 1974, Log Interpretation, vol. II/Applications, p. 49.

*A Mark of Schlumberger

$$\text{Vsh (in\%)} = \frac{\phi\,N}{\phi\,N\ \text{clay}}$$

Where: $\phi\,N$ = neutron porosity of formation
$\phi\,N$ clay = neutron porosity of shale

Schlumberger, 1974, Log Interpretation, vol. II/Applications, p. 50.

In order to determine more accurate lithologies from log data, however, the cross-plot methods (i.e. M-N* plot and MID* plot) should be employed.

M-N* Cross-Plot

The M-N* cross-plot requires, in addition to the neutron and density logs, a sonic log. The sonic log is a porosity log that measures the interval transit time (Δt) which is the reciprocal of the velocity of a compressional sound wave (see: *Schlumberger Log Interpretation Manual, Volume I, Principles,* (1972, p. 37–41). The data combined from all three of these logs are necessary to calculate the lithology dependent quantities M* and N*. M* and N* are essentially independent of primary porosity; therefore, a cross-plot of these two variables makes lithology more apparent.

M* and N* are defined by the following equations:

$$M^* = \frac{\Delta tf - \Delta t}{\rho b - \rho f} \times 0.01$$

$$N^* = \frac{\phi nf - \phi n}{\rho b - \rho f}$$

Where: Δtf = interval transit time of the fluid (189 fresh muds and 185 salt muds)
Δt = interval transit time from the log
ρf = density of the fluid (1.0 fresh muds and 1.1 salt muds)
ρb = bulk density from the log
ϕnf = neutron porosity fluid (use 1.0)
ϕn = neutron porosity from the log. This neutron porosity can be determined from a CNL* log or a sidewall neutron porosity log (SNP*).

*A Mark of Schlumberger

Schlumberger, 1972, Log Interpretation Manual/vol. I, Principles, p. 73.

If the matrix parameters (Δtma, ρma, ϕnma; Table 1) are used instead of log parameters, M* and N* values are calculated for the various minerals (Table 2). These values can then be plotted to determine end-points for the M-N* plot (Fig. 7-2).

Figure 7-2 illustrates a M-N* cross-plot of data from 11,870 feet to 11,900 ft. (Ordovician Red River C-zone) in the Alpar Resources Federal 1-10 Richland County, Montana (Fig. 7-2). The M* and N*

TABLE 1

MATRIX AND FLUID COEFFICIENTS OF SEVERAL MINERALS
AND TYPES OF POROSITY (LIQUID-FILLED BOREHOLES)

	Δtma	ρma	(ΦSNP*)ma	(ϕCNL*)ma
Sandstone (1) (Vma=18,000), $\phi > 10\%$	55.5	2.65	−0.035**	−0.05**
Sandstone (2) (Vma=19,500) $\phi < 10\%$	51.2	2.65	−0.035**	−0.005
Limestone	47.5	2.71	0.00	0.00
Dolomite (1) (ϕ=5.5% to 30%)	43.5	2.87	0.035**	0.085**
Dolomite (2) (ϕ=1.5% to 5.5% & > 30%)	43.5	2.87	0.02**	0.065**
Dolomite (3) (ϕ−0.0% to 1.5%)	43.5	2.87	0.005**	0.04**
Anhydrite	50.0	2.98	−0.005	−0.002
Gypsum	52.0	2.35	0.49***	
Salt	67.0	2.03	0.04	−0.01

From Schlumberger Log Interpretation Manual/Principles, Courtesy Schlumberger Well Services, © 1972, Schlumberger

*A Mark of Schlumberger
**Average values
***Based on hydrogen-index computation

*A Mark of Schlumberger

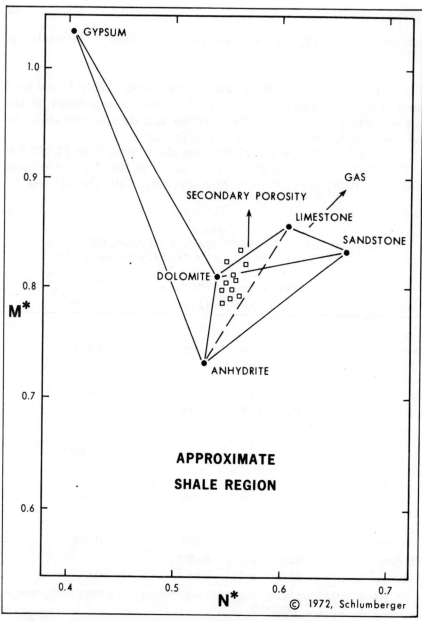

FIG. 7-2: M-N* cross-plot of data from Ordovician Red River C-zone (11,870 ft. to 11,900 ft.) in the Alpar Resources Federal 1-10, Sec. 10, T26N-R55E, Richland County, Montana. From Schlumberger Log Interpretation Manual/Principles, 1972, p. 74. Courtesy of Schlumberger Well Services © 1972, Schlumberger.

*A Mark of Schlumber

TABLE 2
VALUES OF M* AND N* FOR COMMON MINERALS

MINERAL	Fresh Mud (ρ=1.0) M*	N*	Salt Mud (ρ=1.1) M*	N*
Sandstone (1) Vm=18,000	.810	.628	.835	.669
Sandstone (2) Vm=19,500	.835	.628	.862	.669
Limestone	.827	.585	.854	.621
Dolomite (1) ϕ=5.5-30%	.778	.516	.800	.544
Dolomite (2) ϕ=1.5-5.5%	.778	.524	.800	.554
Dolomite (3) ϕ=0-1.5%	.778	.532	.800	.561
Anhydrate ρma=2.98	.702	.505	.718	.532
Gypsum	1.015	.378	1.064	.408
Salt			1.269	1.032

From Schlumberger Log Interpretation Manual/Principles, Courtesy Schlumberger Well Services. © 1972, Schlumberger
*A Mark of Schlumberger

values from this interval cluster in the triangle defined by the end-members limestone, dolomite, and anhydrite. The lithology, therefore, can be classified as an anhydritic limey dolomite (Fig. 7-2). The data points that plot above the dolomite limestone line indicate secondary porosity (i.e. vugs and/or fractures). An advantage of the M-N* plot technique is that M-N* plots can be easily computerized (Bond, 1977).

MID* Cross-plot

A second cross-plot method for identifying lithology and secondary porosity is the MID* (Matrix Identification) plot. Like the M-N* plot, the MID* plot requires: neutron, density, and sonic logs. The first step needed to construct a MID* plot is to determine values for the apparent matrix parameters (ρma)a and (Δtma)a using the

*A Mark of Schlumberger

appropriate neutron-density and sonic-density cross-plot charts. These cross-plot charts, along with instructions on their application, can be obtained from the *Schlumberger Log Interpretation Manual, Volume II, Applications,* (1974, p. 24–27).

The values of apparent matrix density (ρma)a and apparent interval transit time (Δtma)a are then plotted on the MID* plot chart (Fig. 7-3). The data illustrated in Figure 7-3 are from the Alpar Resources Federal 1-10, Richland County, Montana taken over the same interval in the Ordovician Red River C-zone (11,870-11,900) feet) illustrated in the M-N* plot (Fig. 7-2). The values cluster in a triangle defined by the end-members calcite, dolomite, and anhydrite indicating that the lithology is an anhydritic limey dolomite (Fig. 7-3). The data points that fall outside the line between dolomite and limestone show there is some secondary porosity (Fig. 7-3).

All of the methods described are useful in subsurface facies analysis. It has, however, been the writer's experience that the combination gamma ray CNL*-FDC* log and carbonate-shale ratios calculated from the gamma ray, spontaneous potential, or neutron logs have the broadest application. This is the case because of the ease with which mapping data can be obtained, and because many wells have an incomplete log package (i.e. no sonic log available).

Carbonate Rock Type Identification

A significant contribution to log analysis work in recent years has been the attempt to establish relationships between log response and carbonate rock type so that depositional environments can be reconstructed. Cross-plots are used by Pickett (1977) to identify these log response-rock type relationships. Table 3 is a list of various cross-plots employed by Pickett (1977). Pickett's method, however, can be used only when cores or cuttings from selected wells are available. Petrographic analysis of these cores or cuttings is necessary in order to firmly establish rock type. Log responses from the control wells (i.e. wells with petrographic analysis) are cross-plotted and then areas are outlined where the different rock types cluster. Once the clusters are established, log responses from wells without cores or cuttings can be cross-plotted on the chart and their carbonate rock type and depositional environment determined (Fig. 7-4).

The Pickett cross-plot method can be applied to proximity facies

*A Mark of Schlumberger

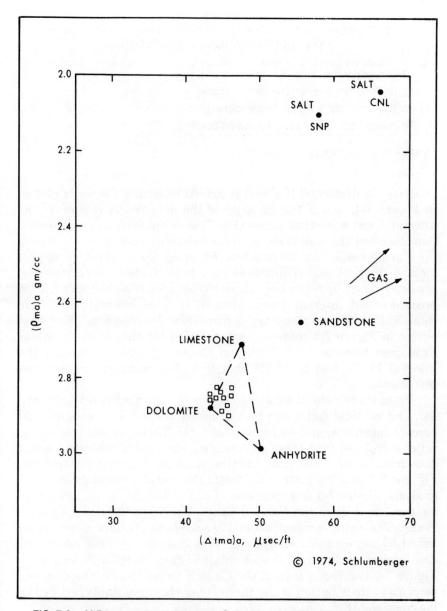

FIG. 7-3: MID* cross-plot of data from Ordovician Red River C-zone (11,870 ft. to 11,900 ft.) in the Alpar Resources Federal 1-10, Sec. 10, T26N-R55E, Richland County, Montana. From Schlumberger Log Interpretation Manual/Applications, 1974, p. 28. Courtesy of Schlumberger Well Services © 1974, Schlumberger.

*A Mark of Schlumberger

TABLE 3
DIFFERENT TYPES OF PICKETT (1977) CARBONATE ROCK
TYPE IDENTIFICATION CROSS PLOTS

Δt (interval transit time) vs. ϕn (neutron porosity)
ρb (bulk density) vs. ϕn (neutron porosity)
ρb (bulk density) vs. Δt (interval transit time)
Rt (deep Laterolog*) vs. ϕn (neutron porosity)

*A Mark of Schlumberger

analysis. To illustrate: If a well is drilled offsetting the wells plotted in Figure 7-4, and if the majority of the data points cluster in the intertidal and supratidal areas (Fig. 7-4), a conclusion can be drawn that the offsetting well is closer to the paleoshoreline.

Facies maps can be constructed using the Pickett cross-plot method. Percentages of different rock type clusters (environments) are determined, or ratios are calculated between two clusters within a predetermined interval (example: a 50-ft. slice interval), and from these values, facies maps are constructed. To illustrate: The open circles in Figure 7-4 represent cross-plotted log data from the Alpar Resources Federal 1-10, Richland County, Montana through the interval 11,800 feet to 11,850 feet (Fig. 7-1) measured in five foot increments.

From these data, the following percentages of supratidal, intertidal, and subtidal facies can be calculated: 30 percent supratidal, 30 percent intertidal, and 40 percent subtidal. These percentages or the ratios calculated from these percentages, when combined with similar data from other wells, can then be used as mapping parameters. Figures 7-5 and 7-6 further illustrate the use of cross-plots in facies analysis. Figure 7-5 is a cross-plot of sonic porosity versus resistivity for the Lower Permian Council Grove B-zone in Ochiltree County, Texas. The oolite grainstone and wackestone clusters were established by petrographic analysis of cores and cuttings; the additional data plotted in Figure 7-5 was obtained from wells with logs only. Figure 7-6 is a facies map of the Council Grove B-zone based on the percentage distribution of the three facies clusters established by the log cross-plot (Fig. 7-5).

In cross-plot facies mapping, if deep resistivity logs are used from both hydrocarbon bearing and water saturated formations, Ro (resistivity of a formation 100% water saturated) can be used in the cross-

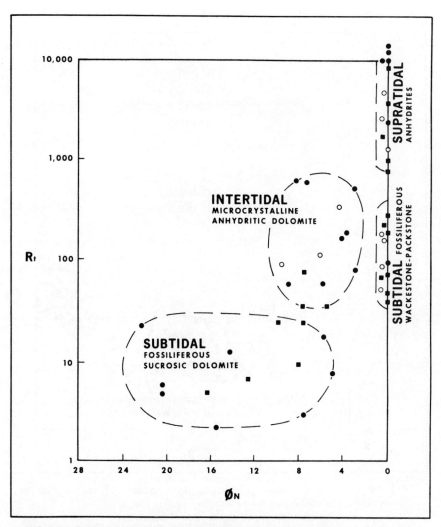

FIG. 7-4: R$_t$ (deep Laterolog*) versus ϕn (neutron porosity) cross plot of data from the Ordovician Red River C and D zones in the Pennzoil No. 1 Norby (squares), Sec. 14, T25N-R57E Richland County, Montana and the Anadarko Production Alpar State 1-28A (solid circles), Sec. 28, T29N-R55E Roosevelt County, Montana. Pennzoil No. 1 Norby data based on thin section analysis of cores and the Anadarko Production Alpar State 1-28A data based on both cores and cuttings. The open circles represent cross-plotted log data from the Alpar Resources Federal 1-10 Sec. 10, T26N-R55E Richland County, Montana through the interval 11,800 ft. to 11,850 ft.

*A Mark of Schlumberger

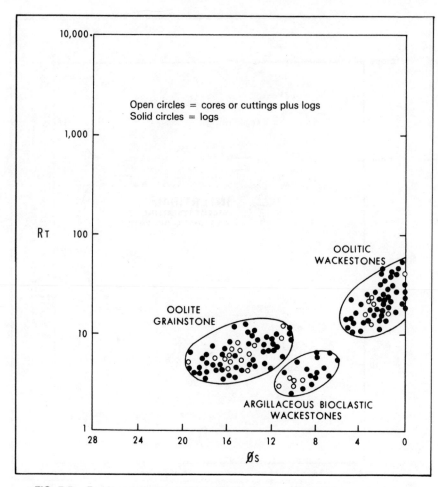

FIG. 7-5: R_t (deep induction log) versus ϕ_s (sonic porosity**) cross-plot from the Lower Permian Council Grove B-zone Ochiltree County, Texas.
**Sonic porosity based on limestone matrix ($\Delta t = 47.6 \, \mu sec/ft.$).

plot rather than the deep resistivity (i.e. R_{ILD}* or R_{LLD}*) value. The formula for calculating R_0 is stated as follows:

$$R_o = 1.0/\phi^2 \times R_w$$

*A Mark of Schlumberger

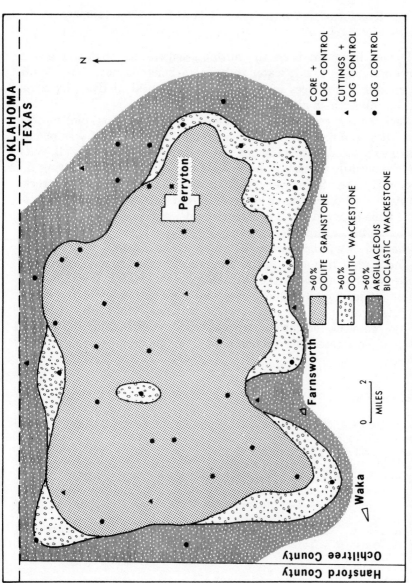

FIG. 7-6: Facies map of the Lower Permian Council Grove B-zone Ochiltree Country, Texas. Facies distribution based on facies clusters established from cross-plotted log data (see: Fig. 7-5).

Where:

ϕ = porosity
R_w = formation water resistivity at formation temperature

It has, however, been the author's experience that changes in reservoir fluids (i.e. salt water to hydrocarbons) do not make significant enough differences to negate environmental analysis by crossplotting unless the hydrocarbon is gas.

Perhaps the greatest advantage of the Pickett cross-plot method is that it maximizes the use of available information. Cores or cuttings are required from a few control wells rather than from all wells. This is especially important because of the difficulty in acquiring the cores or cuttings from every well in an area. Also, since the rocks from each well do not have to be analyzed petrographically, a great deal of time is saved when constructing depositional environments. It is probable that as a larger number of geologists gain familiarity with this method, it will be more widely used in the future.

The author has applied the cross-plot method to subsurface carbonate facies analysis in the Williston Basin and has found it to be reliable. It is, however, most essential for the reader to remember that petrographic analysis of cores or cuttings from control wells in any new zone or formation is necessary to insure that the correct cross-plot parameters are determined.

REFERENCES

Bond, R. C., 1977, M-N*Computer cross-plot for the interpretation of subsurface lithologies: *The Compass, v. 50, no. 2,* p. 36–41.

Pickett, G. R., 1977, Recognition of environments and carbonate rock type identification *in* Formation Evaluation Manual Unit II, section exploration wells: *Oil and Gas Consultants International Inc.,* p. 4–25.

Schlumberger, 1972, *Log Interpretation Manual vol. I,* Principles, 113 p.

——————————, 1974, *Log Interpretation Manual, vol. II,* Applications, 116 p.

Wermund, E. G., 1975, Upper Pennsylvanian Limestone Banks, North Central Texas: Univ. Texas, Austin, *Bur.* Econ. Geology, *Circ.* 75–3, 34 p.

*A Mark of Schlumberger

Index

117

*A mark of Schlumberger

*A mark of Schlumberger